Adaptation, Learning, and Optimization 13

Series Editors-in-Chief

Meng-Hiot Lim
Division of Circuits and Systems
School of Electrical & Electronic
Engineering
Nanyang Technological University
Singapore 639798
E-mail: emhlim@ntu.edu.sg

Yew-Soon Ong
School of Computer Engineering
Nanyang Technological University
Block N4, 2b-39
Nanyang Avenue
Singapore 639798
E-mail: asysong@ntu.edu.sg

T0189576

For further volumes:
http://www.springer.com/series/8335

Shengli Wu

Data Fusion in Information Retrieval

 Springer

Author
Shengli Wu
University of Ulster
United Kingdom

ISSN 1867-4534   e-ISSN 1867-4542
ISBN 978-3-642-44801-0   ISBN 978-3-642-28866-1 (eBook)
DOI 10.1007/978-3-642-28866-1
Springer Heidelberg New York Dordrecht London

Printed on acid-free paper

Springer is part of Springer Science+Business Media (www.springer.com)

Preface

These days the technique of data fusion has been used extensively in many different application areas and information retrieval is one of them. As early as 1980's, researchers began to investigate combining retrieval results generated from many different ways. Since then the problem of data fusion has been investigated or applied by many researchers in the information retrieval community. Looking at the technical implementation of information retrieval systems, one may find that data fusion has been used by many of them, if not all of them, in one way or another. This demonstrates that data fusion, as a useful technique for performance improvement, has been accepted by the information retrieval community. But sometimes, the best possible fusion technique has not been used.

In 2001 I took part in an EU Project that was relevant to data fusion. I published my first paper on data fusion in 2002. Since then, I have been spending most of my research time on data fusion and some related topics in information retrieval. It has been 10 years and I have done a lot of research work on it. I think it is a good idea to put all my research output together in a systematic style so as to make it easy for others to find relevant material. The book should be useful for researchers and students in the areas of information retrieval, web search, digital libraries, and online information service, to obtain the up-to-date data fusion technique. The book may also be useful for people in other research areas such as machine learning, artificial neural networks, classification of text documents or/and multimedia files, and so on, since the data fusion technique has been widely used in those areas as well.

The book has a good coverage of different aspects of the data fusion technique: from retrieval evaluation (Chapter 2) to score normalization (Chapter 3); from score-based methods (Chapters 4, 5, and 6) to ranking-based methods (Chapter 7); from theoretical framework (Chapter 6) to data fusion applications (Chapter 9). Experiments have been conducted and reported for almost all of the aspects discussed. A related problem, fusing results from overlapping databases, is also addressed (Chapter 8).

Information retrieval is an area that requires effort on both theoretical and empirical sides. There is no exception for the data fusion problem. The geometric framework discussed in Chapter 6 provides us a good platform to understand the properties of (score-based) data fusion algorithms; while results of various experiments help us to find useful information such as the relation between rank and relevance in a resultant list, or the performance of a given data fusion algorithm in various kinds of situations.

Apart from some examples of application in Chapter 9, I wrote programs for almost all those empirical studies included in this book. Sometimes it is a bitter experience when I found some results obtained were inconsistent or suspicious, but could not find where the things might go wrong. But after so many data fusion experiments, I am now happy to see that in general, the results are consistent and convincing across different experiments. For example, CombSum, CombMNZ, the Borda count, the linear combination method have been tested in different data sets for many times, and the experimental results are very consistent.

Finally, I would give my thanks to my colleagues and friends, especially Fabio Crestani, Sally McClean, Yaxin Bi, Xiaoqin Zeng, and others, who worked with me on the data fusion problem or helped me somehow; I would give my thanks to my wife, Jieyu Li, who always backs me; and I would give my thanks to my son, Zao Wu, who proof-read most of my published papers and the manuscript of this book as well.

Shengli Wu
University of Ulster
Jordanstown, Northern Ireleand, UK
January, 2012

Contents

1 Introduction 1
   1.1 The Rationale Behind Data Fusion 1
   1.2 Several Typical Data Fusion Methods 2
   1.3 Several Issues in Data Fusion 4

2 Evaluation of Retrieval Results 7
   2.1 Binary Relevance Judgment 7
   2.2 Incomplete Relevance Judgment 10
   2.3 Graded Relevance Judgment 11
   2.4 Score-Based Metrics 14
   2.5 Uncertainty of Data Fusion Methods on Effectiveness
        Improvement 18

3 Score Normalization 19
   3.1 Linear Score Normalization Methods 20
        3.1.1 The Zero-One Linear Method 20
        3.1.2 The Fitting Method 21
        3.1.3 Normalizing Scores over a Group of Queries 22
        3.1.4 Z-Scores 22
        3.1.5 Sum-to-One 23
        3.1.6 Comparison of Four Methods 24
   3.2 Nonlinear Score Normalization Methods 27
        3.2.1 The Normal-Exponential Mixture Model 27
        3.2.2 The CDF-Based Method 27
        3.2.3 Bayesian Inference-Based Method 28
   3.3 Transforming Ranking Information into Scores 29
        3.3.1 Observing from the Data 29
        3.3.2 The Borda Count Model 30
        3.3.3 The Reciprocal Function of Rank 31
        3.3.4 The Logistic Model 31

      3.3.5   The Cubic Model 32

      3.3.6   The Informetric Distribution 33

      3.3.7   Empirical Investigation 35

  3.4   Mixed Normalization Methods 40

  3.5   Related Issues 41

      3.5.1   Weights Assignment and Normalization 41

      3.5.2   Other Relevance Judgments 41

4   Observations and Analyses 43

  4.1   Prior Observations and Points of View 43

  4.2   Performance Prediction of CombSum and CombMNZ........... 46

      4.2.1   Data Fusion Performance 48

      4.2.2   Improvement over Average Performance................. 51

      4.2.3   Performance Improvement over Best Performance 52

      4.2.4   Performance Prediction 54

      4.2.5   The Predictive Ability of Variables 55

      4.2.6   Some Further Observations 57

      4.2.7   Summary 61

  4.3   Comparison of CombSum and CombMNZ 62

      4.3.1   Applying Statistical Principles to Data Fusion 63

      4.3.2   Further Discussion about CombSum and CombMNZ 64

      4.3.3   Lee's Experiment Revisited........................... 66

  4.4   Performance Prediction of the Linear Combination Method 70

5   The Linear Combination Method 73

  5.1   Performance Related Weighting 73

      5.1.1   Empirical Investigation of Appropriate Weights 75

      5.1.2   Other Features..................................... 81

  5.2   Consideration of Uneven Similarity among Results 87

      5.2.1   Correlation Methods 1 and 2........................... 87

      5.2.2   Correlation Methods 3 and 4........................... 88

      5.2.3   Correlation Methods 5 and 6........................... 89

      5.2.4   Empirical Study 91

      5.2.5   Some More Observations 94

  5.3   Combined Weights 97

      5.3.1   Theory of Stratified Sampling and Its Application 97

      5.3.2   Experiments 99

  5.4   Deciding Weights by Multiple Linear Regression 102

      5.4.1   Empirical Investigation Setting 103

      5.4.2   Experiment 1...................................... 104

      5.4.3   Experiment 2...................................... 109

      5.4.4   Experiment 3...................................... 111

  5.5   Optimization Methods 114

  5.6   Summary 116

6   A Geometric Framework for Data Fusion 117
   6.1   Relevance Score-Based Data Fusion Methods
       and Ranking-Based Metrics 117
   6.2   The Geometric Framework 119
   6.3   The Centroid-Based Data Fusion Method 120
   6.4   The Linear Combination Method 126
   6.5   Relation between the Euclidean Distance and Ranking-Based
       Measures 129
   6.6   Conclusive Remarks 132

7   Ranking-Based Fusion 135
   7.1   Borda Count and Condorcet Voting 135
   7.2   Weights Training for Weighted Condorcet Voting 137
   7.3   Experimental Settings and Methodologies 139
   7.4   Experimental Results of Fusion Performance 140
   7.5   Positive Evidence in Support of the Hypothesis 144

8   Fusing Results from Overlapping Databases 149
   8.1   Introduction to Federated Search 149
   8.2   Resource Selection 152
       8.2.1   A Basic Resource Selection Model 152
       8.2.2   Solution to the Basic Resource Selection Model 156
       8.2.3   Some Variants of the Basic Model and Related Solutions ... 159
       8.2.4   Experiments and Experimental Results 163
       8.2.5   Further Discussion 165
   8.3   Results Merging 166
       8.3.1   Several Result Merging Methods 167
       8.3.2   Evaluation of Result Merging Methods 169
       8.3.3   Overlap between Different Results 177
       8.3.4   Conclusive Remarks 178

9   Application of the Data Fusion Technique 181
   9.1   Ranking Information Retrieval Systems with Partial Relevance
       Judgment 181
       9.1.1   How Is the Problem Raised? 182
       9.1.2   Other Options Than Averaging All the Values for Ranking
            Retrieval Systems 185
       9.1.3   Evaluation of the Four Ranking Methods 186
       9.1.4   Remarks 189
   9.2   Ranking Information Retrieval Systems without Relevance
       Judgment 189
       9.2.1   Methodology 191
       9.2.2   Experimental Settings and Results 192

9.3   Applying the Data Fusion Technique to Blog Opinion Retrieval 193
       9.3.1   Data Set and Experimental Settings 194
       9.3.2   Experimental Results 195
       9.3.3   Discussion and Further Analysis 202
9.4   TREC Systems via Combining Features and/or Components 203
       9.4.1   A Hierarchical Relevance Retrieval Model for Entity
               Ranking 203
       9.4.2   Combination of Evidence for Effective Web Search 205
       9.4.3   Combining Candidate and Document Models for Expert
               Search 206
       9.4.4   Combining Lexicon-Based Methods to Detect Opinionated
               Blogs 207
       9.4.5   Combining Resources to Find Answers to Biomedical
               Questions.. 210
       9.4.6   Access to Legal Documents: Exact Match, Best Match,
               and Combinations..................................... 210
       9.4.7   Summary 212

References 213

A   Systems Used in Section 5.1 221

B   Systems Used in Section 5.2.4 225

C   Systems Used in Sections 7.3-7.5 227

Notation used throughout this book (1)

Symbol	Meaning
CombMNZ	a data fusion method
CombSum	a data fusion method
D	document collection
$|D|$	number of documents in D
d_i	a document in collection D
$dist(A,B)$	Euclidean distance between A and B
$g(d_i)$	global score of document d_i, used for ranking the fused result
IR	a group of information retrieval systems
ir_i	an information retrieval system
LC	linear combination, a data fusion method
LCP	the LC method with performance-level weighting
LCP2	the LC method with performance square weighting
LCR	the LC method with weights trained by multiple regression
L_i	a result (a ranked list of documents) from ir_i for given D and q
$|L_i|$	number of documents in L_i
o_rate	overlap rate between two or more retrieval results
p_i	performance of a retrieval system ir_i over one or a group of queries
$Pr(t)$	probability of a document at rank t being relevant
Q	a group of queries
$|Q|$	number of queries in Q
q or q_i	a query in Q
r_i	raw score of document d_i
s_i	normalized score of document d_i
t_i	ranking position of document d_i
w_i	weight assigned to ir_i

Notation used throughout this book (2)

Metric	Meaning
AP	average precision over all relevant documents, or average precision for short
APD	average precision over all documents
bpref	binary preference
DCG	discounted cumulative gain
infAP	inferred average precision
NAPD	normalized average precision over all documents
NDCG	normalized discounted cumulative gain
NDCG@k	NDCG that sets 1 to be the weight of the first k documents
P@x	precision at x document-level, a metric for retrieval evaluation
PIA	percentage of improvement of the fused result over average of all component results
PIA(AP)	PIA measured by AP
PIA(RP)	PIA measured by RP
PIA(RR)	PIA measured by RR
PIB	percentage of improvement of the fused result over the best component result
PIB(AP)	PIB measured by AP
PIB(RP)	PIB measured by RP
PIB(RR)	PIB measured by RR
PFB	percentage of runs that the fused result is better than the best component result
PFB(AP)	PFB measured by AP
PFB(RP)	PFB measured by RP
PFB(RR)	PFB measured by RR
RP	recall-level precision
RR	reciprocal rank

Chapter 1
Introduction

Data fusion has been used in many different application areas. However, the data fusion technique in information retrieval is special mainly due to its result presentation style – a ranked list of documents. In this introductory part of the book, we discuss several typical data fusion methods in information retrieval, including CombSum, CombMNZ, the linear combination method, the Borda count, the Condorcet voting, and others. At the end of this section, an introductory remark is given for several major issues of data fusion in information retrieval.

1.1 The Rationale Behind Data Fusion

In recently years, data fusion has been applied to many different application areas such as neural networks, classification, multi-sensor, image, and so on, to improve system performance. Some other terms such as combination, assembling, aggregation, are also used by some people to refer to the same or similar ideas. In information retrieval, data fusion is also known as meta-search.

The key point in information retrieval is how to rank all the documents that are retrieved. This is the task of the ranking algorithm. For different information retrieval systems, the difference between them can be in many aspects:

- Many different models (for example, the boolean model, the vector space model, the probabilistic model, the language model, and many others) can be used.
- Different treatments on many other aspects such as parsing rules, stemming, phrase processing, relevance feedback techniques, and so on.
- Documents and queries can be represented in different ways.
- For any individual information retrieval system, different combinations of those options may be taken and very often many of the developed systems are competitive in effectiveness.

Because of these, fusing the results from multiple retrieval systems (features, components, and so on) so as to obtain more effective results is a desirable approach.

S. Wu: Data Fusion in Information Retrieval, ALO 13, pp. 1–5.
springerlink.com   © Springer-Verlag Berlin Heidelberg 2012

Previous investigations demonstrate that better result is achievable if appropriate data fusion methods are used.

1.2 Several Typical Data Fusion Methods

Suppose there is a document collection D and a group of retrieval systems $IR = \{ir_i\}$ for ($1 \leq i \leq n$). All retrieval systems ir_i ($1 \leq i \leq n$) search D for a given query q and each of them provides a ranked list of documents $L_i = <d_{i1}, d_{i2}, ..., d_{im}>$. Sometimes a relevance score $s_i(d_{ij})$ is associated with each of the documents in the list. The data fusion technique is to use some algorithms to merge these n ranked lists into one. We hope that the fused result is more effective than those component results.

CombSum [32, 33] uses the following equation

$$g(d) = \sum_{i=1}^{n} s_i(d) \tag{1.1}$$

to calculate scores for every document d. Here $s_i(d)$ is the score that ir_i assigns to d. If d does not appear in any L_i, then a default score (e.g., 0) must be assigned to it. After that, every document d obtain a global score $g(d)$ and all the documents can be ranked according to the global scores they obtain.

Another method CombMNZ [32, 33] uses the equation

$$g(d) = m * \sum_{i=1}^{n} s_i(d) \tag{1.2}$$

to calculate scores. Here m is the number of results in which document d appears.

The linear combination method [98, 99, 112, 113] uses the equation below

$$g(d) = \sum_{i=1}^{n} w_i * s(d) \tag{1.3}$$

to calculate scores. w_i is the weight assigned to system ir_i. The linear combination method is very flexible since different weights can be assigned to different information retrieval systems. Obviously, the linear combination is a general form of CombSum. If all the weights w_i are equals to 1, then the linear combination is the same as CombSum. Note that how to assign weights to different retrieval systems is an important issue. This will be addressed later (mainly in Chapter 5) in this book. Now let us see an example to illustrate how these methods work.

Example 1.2.1. Suppose for a document collection D and a given query q, three information retrieval systems search D and provide three ranked lists of documents. $L_1 = <(d_1, 0.9), (d_2, 0.7), (d_3, 0.4)>$, $L_2 = <(d_1, 0.8), (d_3, 0.6), (d_4, 0.4)>$, $L_3 = <(d_4, 0.8), (d_2, 0.7), (d_3, 0.3)>$. Here each document retrieved is assigned an estimated relevance score and documents rank 1, 2, and so on, from left to right. What are the fusion results if we use CombSum, CombMNZ, and the linear combination method to fuse them?

For CombSum, $s(d_1) = 0.9 + 0.8 = 1.7$; $s(d_2) = 0.7 + 0.7 = 1.4$; $s(d_3) = 0.4 + 0.6 + 0.3 = 1.3$; $s(d_4) = 0.4 + 0.8 = 1.2$. Therefore, the final ranking is F(CombSum) = $<(d_1, 1.7), (d_2, 1.4), (d_3, 1.3), (d_4, 1.2)>$, or $<(d_1, 1.7), (d_2, 1.4), (d_3, 1.3)>$, if one wishes to provide only the top three documents as all component results do.

For CombMNZ, $s(d_1) = (0.9 + 0.8) *2 = 3.4$; $s(d_2) = (0.7 + 0.7) * 2 = 2.8$; $s(d_3) = (0.4 + 0.6 + 0.3) *3 = 3.9$; $s(d_4) = (0.4 + 0.8) *2 = 2.4$. Therefore, the final ranking is F(CombMNZ) = $<(d_3, 3.9), (d_1, 3.4), (d_2, 2.8), (d_4, 2.4)>$.

If 1, 2, and 4 are assigned as weights to L_1, L_2, and L_3, respectively, and we use the linear combination (LN) to fuse them, then $s(d_1) = 0.9 * 1 + 0.8 * 2 = 2.5$; $s(d_2) = 0.7 * 1 + 0.7 * 4 = 3.5$; $s(d_3) = 0.4 * 1 + 0.6 * 2 + 0.3 * 4 = 2.8$; $s(d_4) = 0.4 * 2 + 0.8 * 4 = 4.0$. Therefore, the final ranking is F(LN) = $<(d_4, 4.0), (d_2, 3.5), (d_3, 2.8), (d_1, 2.5)>$. $\qquad\square$

The Borda count [3] is an election method in which voters rank candidates in order of preference. It can be used for data fusion in information retrieval if we regard documents as candidates and information retrieval systems as voters. For any ranked list of m documents, the first one is assigned a score of m, the second one is assigned a score of $m - 1$, and so on, and the last document is assigned a score of 1. Then the same method as CombSum can be used for the fusion process. The difference between the Borda count and CombSum is how to obtain relevance scores for the documents involved. The Borda count generates such scores from ranking information, while CombSum uses the scores provided by the information retrieval systems themselves. The Borda count is useful if scores are of a poor quality or not available at all. The Borda count can be easily extended to be weighted Borda count by using linear combination to replace CombSum for the fusion process.

The Condorcet voting [61] is another election method. It is a pair-wise voting, i.e., it compares every possible pair of candidates to decide the preference of them. Suppose there are m candidates (documents) and n voters (information retrieval systems), and each of the voters provides preference for every pair of candidates. A matrix can be used to present the competition process. Every candidate (document) appears in the matrix as a row and as a column as well. If there are m candidates (documents), then we need m^2 elements in the matrix in total. Initially 0 is written to all the elements. If d_i is preferred to d_j, then we add 1 to the element at row i and column j. This needs to be done over and over until all the information retrieval systems have been processed. Next for each element a_{ij}, if $a > n/2$, then d_i beats d_j; if $a < n/2$, then d_j beats d_i; otherwise ($a = n/2$), there is a draw between d_i and d_j. The total score of each candidate (document) is worked out by summarizing the scores it obtains in all pair-wise competitions. Finally the ranking is achievable based on the total scores calculated. Let us take an example to see how the Condorcet voting can work as a data fusion method in information retrieval.

Example 1.2.2. let us assume that $L_1 = \ <d_2, d_3, d_1, d_4>$. $L_2 = \ <d_3, d_4, d_1, d_2>$, and $L_3 = \ <d_1, d_3, d_2, d_4>$ Now we use the Condorcet voting to fuse them. In L_1, d_2 has higher preference than d_3, d_1, and d_4; d_3 has higher preference than d_1 and

d_4; and d_1 has higher preference than d_4. We add 1 to the corresponding units and the matrix looks like this:

R		Opponent			
u		d_1	d_2	d_3	d_4
n	d_1	-	0	0	1
n	d_2	1	-	1	1
e	d_3	1	0	-	1
r	d_4	0	0	0	-

We continue processing with L_2 and L_3 and the matrix is as follows:

R		Opponent				Total Scores
u		d_1	d_2	d_3	d_4	
n	d_1	-	2	1	2	2
n	d_2	1	-	1	2	1
e	d_3	2	2	-	3	3
r	d_4	1	1	0	-	0

Note that there are 3 voters (information retrieval systems) in total. For each element (a_{ij}) in the matrix, if a_{ij} is 2 or above, then d_i defeats d_j; if a_{ij} is 1 or less, then d_j defeats d_i. For the document in each row, we count how many times it wins over other documents. The total number of wins is written down to the right side of the above matrix. The final fused ranking is $<(d_3, 3), (d_1, 2), (d_2, 1), (d_4, 0)>$.

Different weights can be set for different information retrieval systems to extend the Condorcet voting to become weighted Condorcet voting. The only change for the fusion process is, for any pair $<d_i, d_j>$ in L_k, w_k (the weight assigned to L_k), rather than 1, is added to the unit at row d_i and column d_j. The rule for deciding the winner/loser of the voting is: for any element a_{ij}, if $a_{ij} > \sum_{i=1}^n w_i/2$, then d_i beats d_j; if $a_{ij} < \sum_{i=1}^n w_i/2$, then d_j beats d_i; if $a_{ij} = \sum_{i=1}^n w_i/2$, then there is a draw between d_j and d_i [1].

In the above we have discussed a few typical data fusion methods. In the last couple of years more data fusion methods have been proposed. We will discuss some of them later.

1.3   Several Issues in Data Fusion

Several major issues in data fusion include:

- Retrieval evaluation: how to evaluate retrieval results is a key factor that affects the performance of data fusion algorithms. Several different types of evaluation systems are reviewed in Section 2.

---

[1] More detailed discussion about the Condorcet voting and weighted Condorcet voting will be given in Section 7.

- Score normalization: since raw scores from different retrieval systems are usually not comparable, it is required to normalize them in a systematic way to make them comparable. This issue is addressed in Section 3.
- In most cases, uncertainty is a norm for the effectiveness of the fused result if we look at the result individually. However, if we investigate the problem by carrying out large-scale experiments, then it is possible to find out the factors that affect the performance of data fusion significantly. Such issues are addressed in Section 4.
- Data fusion methods in information retrieval can be divided into different categories based on some criteria. According to what information they require, we may divide them into two types: score-based and ranking-based. Another dimension is that in the fusing process we may treat all component systems equally or biased to some of them. Thus either equal weight or different weights can be assigned to different systems for better results. a range of different methods of weight assignment for score-based methods are addressed in Section 5. Furthermore, score-based methods are discussed formally in Section 6 using a geometric framework. Ranking-based methods, including the Condorcet voting and the weighted Condorcet voting, are discussed in Section 7.
- The methods discussed for data fusion in information retrieval have to work on the multiple results from different information retrieval systems with the identical document collection. A slightly different scenario is that multiple information retrieval systems retrieve documents from overlapping document collections. This can be useful, for example, in developing web meta-search engines by fusing results from multiple web search engines. This issue is addressed in Section 8.
- Interestingly, the data fusion methods can be used for different tasks in information retrieval. Some such examples are discussed in Section 9.

Chapter 2
Evaluation of Retrieval Results

Evaluating retrieval results is a key issue for information retrieval systems as well as data fusion methods. One common assumption is that the retrieval result is presented as a ranked list of documents. Under such an assumption [1], we review some retrieval evaluation systems including binary relevance judgment, graded relevance judgment, and incomplete relevance judgment. We also introduce some metrics that will be used later in this book.

In recent years, there are quite a few information retrieval evaluation events such as TREC [2], NCTIR [3], CELF [4]. Some of the methods in this chapter are defined in the context of TREC.

2.1 Binary Relevance Judgment

First let us look at the situation in which binary relevance judgment is used for evaluation. In binary relevance judgment, documents are classified into two categories: relevant or irrelevant. Many different metrics such as average precision, recall-level precision, precision at n (10 is most commonly used, other common options include 5, 20, and 100) document level, normalized discounted cumulative gain, the reciprocal rank, and many others, have been proposed for retrieval evaluation. Almost all those commonly used metrics are ranking-based. It means that the ranking positions of relevant and irrelevant documents are the only concern.

Suppose that D is a collection of documents, q is a given query, and $total\_r$ is the total number of relevant documents in D for q. An information retrieval system ir

---

[1] Other alternative assumptions than a ranked list of documents are possible. They will lead to different evaluation systems. For example, a retrieval result can be presented as a hierarchical structure.

[2] http://trec.nist.gov/

[3] http://research.nii.ac.jp/ntcir/index-en.html

[4] http://www.clef-initiative.eu/

S. Wu: Data Fusion in Information Retrieval, ALO 13, pp. 7–18.
springerlink.com   © Springer-Verlag Berlin Heidelberg 2012

provides a ranked list of documents $L = <d_1, d_2, ..., d_n>$. Average precision over all relevant documents, or average precision (AP) is defined as

$$AP = \frac{1}{total\_r} \sum_{i=1}^{total\_r} \frac{i}{t_i} \qquad (2.1)$$

where t_i is the ranking position of the i-th relevant documents in the resultant list L. If not all $total\_r$ relevant documents appear in the result, then we can use the following equation instead (m is the number of relevant documents that appear in L)

$$AP = \frac{1}{total\_r} \sum_{i=1}^{m} \frac{i}{t_i} \qquad (2.2)$$

Example 2.1.1. There are 4 relevant documents in the whole collection and 2 of them are retrieved in the ranking positions of 2 and 4 in L, then AP(L) = 1/4*(1/2 + 2/4) = 0.25.   □

Normalized average precision over all m documents (NAPD) [122] can de defined as

$$NAPD = \frac{1}{m * napd\_best} \sum_{i=1}^{m} \frac{nu(i)}{i} \qquad (2.3)$$

where $nu(i)$ is a function, which gives the number of relevant documents in the top-i documents. $napd\_best$ is a normalization coefficient, which is the best possible NAPD value for such a resultant list. Note that when calculating the best NAPD value, all relevant documents to the query need to be considered as in the calculation of AP.

Example 2.1.2. There are 4 relevant documents in the whole collection and 2 of them are retrieved in the ranking positions of 2 and 4 in L, there are 5 documents in L, or $|L| = 5$, then $napd\_best$ = (4 + 4/5)/5 = 0.96, NAPD = (0 + 1/2 + 1/3 + 2/4 + 2/5)/(5*0.96) = 0.3611.   □

Recall-level precision (RP) is defined as the percentage of relevant documents in the top $total\_r$ documents in L.

Precision at 10 document level (P@10) is defined as the percentage of relevant documents in the top 10 documents in L.

Normalized discounted cumulative gain (NDCG, or NDCG@k) is introduced in [44]. Each ranking position in a resultant document list is assigned a given weight. The top ranked documents are assigned the heaviest weights since they are the most convenient ones for users to read. A logarithmic function-based weighting schema was proposed in [44], which needs to take a particular whole number k. The first k documents are assigned a weight of 1; then for any document ranked i that is greater than k, its weight is $w(i) = ln(k)/ln(i)$. Considering a resultant document list up to m documents, its discount cumulated gain (DCG, or DCG@k) is defined as

$$DCG = \sum_{i=1}^{m} (w(i) * r(i))$$

where $r(i)$ is defined as: if the i-th document is relevant, then $r(i) = 1$; if the i-th document is irrelevant, then $r(i) = 0$. DCG can be normalized using a normalization coefficient DCG_best, which is the DCG value of the best resultant lists. Therefore, we have:

$$NDCG = \frac{1}{DCG\_best} \sum_{i=1}^{t} (w(i) * r(i)) = \frac{DCG}{DCG\_best} \tag{2.4}$$

Example 2.1.3. In a resultant list L of 10 documents, 4 of them are relevant and ranked at 1, 3, 5, and 8. We also know that there are 6 relevant documents in the whole collection. What are the values of DCG and NDCG of it (set $k = 2$)?

$$DCG@2(L) = \sum_{i=1}^{t} w(i) * r(i) = 1 + \frac{ln2}{ln3} + \frac{ln2}{ln5} + \frac{ln2}{ln8} = 2.3949$$

$$DCG\_best@2 = 1 + 1 + \frac{ln2}{ln3} + \frac{ln2}{ln4} + \frac{ln2}{ln5} + \frac{ln2}{ln6} = 3.9484$$

$$NDCG@2(L) = \frac{DCG(L)}{DCG\_best} = 0.6065 \qquad \square$$

The reciprocal rank (RR) is defined as

$$RR = \frac{1}{f_r} \tag{2.5}$$

where f_r is the ranking position of the first relevant document that appears in the result.

We say that a ranking-based metric is reasonably defined, if it observes the following four rules:

- if there are two relevant documents at different ranks, then exchanging their positions does not affect the effectiveness of the result;
- if there are two irrelevant documents at different ranks, then exchanging their positions does not affect the effectiveness of the result;
- if a relevant document is followed by an irrelevant document, then exchanging their positions degrades the effectiveness of the result. However, some metrics may not be sensitive enough to reflect all such changes. Anyhow, by any reasonably defined metric, the effectiveness of a result L should be at least as effective as L', if L' includes the same group of documents as L and all the documents in L' are in the same rank as in L except one relevant document and its subsequent irrelevant document in L exchange their ranking positions in L';
- if an irrelevant document is followed by a relevant document, then exchanging their positions upgrades the effectiveness of the result. However, some metrics may not be sensitive enough to reflect all such changes. Anyhow, by any reasonably defined metric, the effectiveness of a result L' should be at least as effective as L, if L' includes the same group of documents as L and all the documents in

L' are in the same rank as in L except one irrelevant document and its subsequent relevant document in L exchange their ranking positions in L'.

We can prove that all ranking-based metrics discussed above are reasonably defined.

Example 2.1.4. Decide if average precision (AP) is reasonably defined or not.

In Equation 2.1, there are *total_r* addends in the summation. Let us assume that a relevant document is in ranking position t and an irrelevant document in ranking position $t + 1$. If we swap the positions of these two documents, then one and only one of the addends in the summation is affected ($m/t \longrightarrow m/(t + 1)$) and the AP value is smaller. Likewise, if an irrelevant document is in ranking position t and a relevant document in ranking position $t + 1$ and we swap the positions of those two documents, then one and only one of the addends in the summation is affected ($m/(t + 1) \longrightarrow m/t$) and the AP value is larger. It proves that AP is reasonably defined. It also proves that AP is very sensitive since any swap of relevant and irrelevant documents will affect its value.   □

Example 2.1.5. One metric is defined as average precision of documents that rank between 5 and 10 (P@(5-10)). Decide whether this metric is reasonably defined.

As a matter of fact, this metric is not reasonably defined. We prove this by an example. For a result L, let us assume that the 4th-ranked document is irrelevant and the 5-th ranked document is relevant. Now we swap the two documents and obtain a new result L'. If measured by P@(5-10), L is less effective than L'. This is in contradiction to the third rule of a reasonably defined metric.   □

If a ranking-based metric is not reasonably defined, then it should not be used because of its erratic behaviour.

2.2 Incomplete Relevance Judgment

These days larger and larger document collections are used for retrieval evaluation. It is not affordable to judge all the documents retrieved any more. Therefore, incomplete relevance judgment, or partial relevance judgment, is commonly used. For example, in TREC, A pooling policy is used. That is, for a group of runs submitted to a task for a given topic, only a certain number (say, 100) of top-ranked documents from all or some submitted runs are put into a pool. All the documents in the pool are judged, and all those documents that are not in the pool are not judged and are treated as irrelevant documents.

The TREC's pooling policy does not affect some metrics such as precision at a given cut-off document level. However, in the evaluation of information retrieval systems, both precision and recall are important aspects and many metrics concern both of them at the same time. In order to obtain accurate values for such metrics, complete relevance judgment is required. For example, AP and RP are such metrics and they are used widely.

In the context of TREC, Zobel [132] investigated the reliability of the pooling method. He found that in general the pooling method was reliable, but that recall was overestimated since it was likely that 30% - 50% of the relevant documents had not been found. Therefore, some alternative metrics have been defined for incomplete relevance judgment. bpref [15] is one of them. For a topic with $total\_r$ relevant documents where r is a relevant document and n is a member of the first $total\_r$ judged non-relevant documents as retrieved by the system, It is defined as

$$bpref = \frac{1}{total\_r} \sum_r 1 - \frac{|n\ ranked\ higher\ than\ r|}{total\_r} \tag{2.6}$$

A variant of bpref is bpref@10. It is defined as

$$bpref@10 = \frac{1}{total\_r} \sum_r 1 - \frac{|n\ ranked\ higher\ than\ r|}{10 + total\_r} \tag{2.7}$$

where n is a member of the top $10 + total\_r$ judged non-relevant documents as retrieved by the system.

Next let us see inferred AP (infAP), which is related to AP [4]. This metric is defined in the environment of TREC or in a similar environment where the pooling policy is applied. This metric needs a given number t, so that we can calculate the expected precision above ranking position t. Within the $t - 1$ documents above rank t, there are two main types of documents: documents that are not in the pool ($D_{non-pool}$), which are assumed to be non-relevant, and documents that are within the pool (D_{pool}). For the documents that are within the pool, there are documents that are un-sampled, documents that are sampled and relevant (D_{rel}), and documents that are sampled and irrelevant ($D_{non-rel}$). infAP is defined as

$$infAP = t + \frac{t-1}{t} \{ \frac{|D_{pool}|}{t-1} * \frac{|D_{rel}|}{|D_{rel}| + |D_{non-rel}|} \} \tag{2.8}$$

where $|D_{pool}|$, $|D_{rel}|$, $|D_{non-rel}|$ denotes the number of documents in D_{pool}, D_{rel}, $D_{non-rel}$, respectively. If no documents above k are sampled, then both $|D_{rel}|$ and $|D_{rel}| + |D_{non-rel}|$ become zero. In order to avoid this from happening, we can add a small value ε to both the number of relevant and number of non-relevant documents sampled. Thus, we have

$$infAP = t + \frac{t-1}{t} \{ \frac{|D_{pool}|}{t-1} * \frac{|D_{rel}| + \varepsilon}{|D_{rel}| + |D_{non-rel}| + 2\varepsilon} \} \tag{2.9}$$

2.3 Graded Relevance Judgment

It is straightforward to expand the situation of binary relevance judgment into graded relevance judgment. In graded relevance judgment, documents are divided into $n + 1$

categories: grades $n, n-1, ..., 0$ $(n \geq 2)$. The documents in grade n are the most relevant, which are followed by the documents in grade $n-1, n-2,...,1$, and the documents in grade 0 are irrelevant. One primary assumption we take for these documents in various grades is: any document in grade n is regarded as 100% relevant and 100% useful to users, and any document in grade i $(i < n)$ is regarded as $i/n\%$ relevant and $i/n\%$ useful to users. One natural derivation is that any document in grade i is equal to i/n documents in grade n on usefulness.

NDCG was defined for graded relevance judgment, but many other commonly used metrics, such as average precision and recall-level precision, were defined in the condition of binary relevance judgment. Now let us see how we can generalize them to make them suitable for graded relevance judgment [120]. Suppose there are $n+1$ relevance grades ranging from 0 to n, and each document d_i is assigned a grade $gr(d_i)$ according to its degree of relevance to the given query. Let us assume that, in the whole collection, there are $total\_n$ documents whose grades are above 0, and $total\_n = z_1 + z_2 + ... + z_n$, where z_i denotes the number of documents in grade i.

It is straightforward to generalize P@10 as

$$P@10 = \frac{1}{10*n}\{\sum_{i=1}^{10} gr(d_i)\} \tag{2.10}$$

Before discussing the generalization of other metrics, let us introduce the concept of the best resultant list. For a given information need, a resultant list L is best if it satisfies the following two conditions:

- all the documents whose grades are above 0 appear in the list;
- for any document pair d_i and d_j, if d_i is ranked in front of d_j, then $gr(d_i) \geq gr(d_j)$.

Many resultant lists can be the best at the same time since more than one document can be in the same grade and the documents in the same grade can be ranked in different orders. In any best resultant lists L, a document's grade is only decided by its rank; or for any given rank, a document's grade is fixed. Therefore, we can use $gr\_best(d_j)$ to refer to the grade of the document in ranking position j in one of the best resultant lists. We may also sum up the grades of the documents in top z_n, top $(z_n + z_{n-1})$,..., top $(z_n + z_{n-1} +...+ z_1)$ positions for any of the best resultant lists (these sums are the same for all the best resultant lists):

$$sb_n = \sum_{i=1}^{z_n} gr(d_i)$$

$$sb_{n-1} = \sum_{i=1}^{z_n+z_{n-1}} gr(d_i)$$

$$...$$

$$sb = sb_1 = \sum_{i=1}^{z_n+z_{n-1}+...+z_1} gr(d_i)$$

One simple solution to calculate RP for a resultant list is to use the formula $RP = \frac{1}{sb}\sum_{j=1}^{total\_n} gr(d_j)$. However, this formula does not distinguish positions difference of documents if they are located in top $total\_n$. The effect is the same for any document occurring in ranking position 1 or ranking position $total\_n$. To avoid this drawback, we can use a more sophisticated formula. First we only consider the top z_n documents and use $\frac{1}{sb_n}\sum_{j=1}^{z_n} gr(d_j)$ to evaluate their precision, next we consider the top $z_n + z_{n-1}$ documents and use $\frac{1}{sb_{n-1}}\sum_{j=1}^{z_n+z_{n-1}} gr(d_j)$ to evaluate their precision, continue this process until finally we consider all top $total\_n$ documents by $\frac{1}{sb_1}\sum_{j=1}^{z_n+z_{n-1}+\ldots+z_1} gr(d_j)$. Combining all these, we have

$$RP = \frac{1}{n}\{\frac{1}{sb_n}\sum_{j=1}^{z_n} gr(d_j) + \frac{1}{sb_{n-1}}\sum_{j=1}^{z_n+z_{n-1}} gr(d_j) + \ldots + \frac{1}{sb_1}\sum_{j=1}^{z_n+z_{n-1}+\ldots+z_1} gr(d_j)\}$$

(2.11)

Note that in the above Equation 2.11, each addend inside the braces can vary from 0 to 1. There are n addends. Therefore, the final value of RP calculated is between 0 and 1 inclusive.

Next let us discuss average AP. It can be defined as

$$AP = \frac{1}{total\_n}\sum_{i=1}^{total\_n} \frac{\sum_{j=1}^{i} gr(d_{t_j})}{\sum_{j=1}^{i} gr\_best(d_{t_j})}$$

(2.12)

where t_j is the ranking position of the j-th document whose grade is above 0, $\sum_{j=1}^{i} gr(d_{t_j})$ is the total sum of grades for documents up to rank t_i, and $\sum_{j=1}^{i} gr\_best(d_{t_j})$ is the total sum of grades for documents up to rank t_j in the best result. Considering all these $total\_n$ documents in the whole collection whose grades are above 0, AP needs to calculate the precision at all these document levels ($t_1, t_2,\ldots,$ t_{total_n}). At any t_i, precision is calculated as $\frac{\sum_{j=1}^{i} gr(d_{t_j})}{\sum_{j=1}^{i} gr\_best(d_{t_j})}$, whose value is always in the range of 0 and 1.

Normalized average precision over all documents (NAPD) can be defined as

$$NAPD = \frac{1}{m * napd\_best}\{\sum_{i=1}^{m}\sum_{j=1}^{i} \frac{gr(d_j)}{i}\}$$

(2.13)

$\sum_{j=1}^{i} \frac{gr(d_j)}{i}$ is precision at document level i, and $\frac{1}{m}\{\sum_{i=1}^{m}\sum_{j=1}^{i} \frac{gr(d_j)}{i}\}$ is average precision at all document levels (1 - m), and $1/napd\_best$ is the normalization coefficient. $napd\_best$ is the NAPD value for one of the best resultant lists.

Expected reciprocal rank (ERR) can be defined as in [19]

$$ERR@k = \sum_{i=1}^{k} \frac{T(g_i)}{i}\prod_{j=1}^{i-1}(1 - T(g_i))$$

(2.14)

where $T(g) = \frac{2^g - 1}{16}$ are the relevance grades associated with the top k documents.

Similar to binary relevance judgement, we can set up rules for reasonably defined metrics in the condition of graded relevance judgment. We can also prove that P@10, AP, RP, and so on, as defined above, are reasonably defined.

2.4 Score-Based Metrics

All those metrics defined so far only concern about the positions of relevant and irrelevant documents, therefore, they are referred to as ranking-based metrics. However, in some cases, only a single list of documents as retrieval result is not enough. Let us consider two examples. The first is for the application of patent and legal retrieval [95], it may require that all relevance documents need to be retrieved. The second one is that a researcher wants to find all articles that are relevant to his/her research topic. In both situations, if the retrieval system only provides a large number of ranked documents for a given information need, then the user may feel very frustrated. First, unless the user checks all the documents in the result, one has no idea how many relevant documents are in the whole list. Second, if the user plans to find a given number (say, 10) relevant documents, one has no idea how many documents needs to be checked in the list. Third, one has no idea if all the relevant documents have been included in the result or not.

As a matter of fact, it is a difficult task for existing information retrieval systems and digital libraries to provide a precise answer to the above questions. However, some reasonable estimations, which can be manageable to provide by many information retrieval systems and digital libraries, are also very useful. If a large amount of documents need to be checked, then it is better to know (at least roughly) the number of them before starting to do so. Then an appropriate period of time can be arranged for that.

In order to solve the above problems, more information is needed for retrieval result presentation. One solution is to provide an estimated score for each of the documents retrieved [125]. With associated scores, users can have a clearer view of the result; it is also possible to give a reasonable answer to the above questions.

In some advanced IR applications such as filtering, resource selection and data fusion, people find that relevance scores of documents are highly desirable. Therefore, different kinds of score normalization methods, which map internal retrieval status values to probabilities of relevance, have been proposed and their effectiveness has been investigated in [18, 48, 55, 60, 66].

When ranking documents for a given information need, most information retrieval systems/digital libraries assign a retrieval status value to every document in the document collection, then rank all the documents according to the status values they obtain. It should be noted that such retrieval status values usually do not necessarily have to be the (estimated) relevance probabilities/degrees of the documents. As a consequence, little effort has been made on approximating the relationship between retrieval status values and relevance probabilities/degrees [66]. I advocate that more attention should be paid to this issue.

Suppose that binary relevance judgment is used and scores are explained as estimated probabilities of the documents being relevant to the query q. For the given query q, an information retrieval system ir retrieves a group of documents $L=\{d_1,d_2,...,d_n\}$, each of which (d_i) is assigned a corresponding relevance probability score s_i for $(1 \leq i \leq n)$. Further, we assume that these documents have been ranked according to their scores. Therefore, we have $s_1 \geq s_2 \geq ... \geq s_n$. Now we are in a better position to answer those questions raised at the beginning of this section. First let us discuss how to answer the first question: how many relevant documents are in the list?

Obviously, as a point estimate we can expect $\sum_{i=1}^{n} s_i$ relevant documents in resultant list L. We can also estimate the probabilities that L includes $0,1,...,n$ relevant documents are:

$$Pr(L,0) = \prod_{i=1}^{n}(1 - s_i)$$

$$Pr(L,1) = \sum_{i=1}^{n}[\prod_{j=1 \wedge j \neq i}^{n}(1 - s_j)]s_i$$

$$.........$$

$$Pr(L,k) = \sum_{i=1}^{\frac{n!}{k!(n-k)!}}[(\prod_{j=1}^{k} s_{ij})(\prod_{j=1}^{n-k}(1 - s'_{ij}))]$$

$$.........$$

$$Pr(L,n) = \prod_{i=1}^{n} s_i$$

where $Pr(k)$ $(1 \leq k \leq n)$ is the probability that resultant list L includes just k relevant documents. We explain a little more about $Pr(k)$. For k relevant documents in a list of n $(n > k)$ documents, there are $\frac{n!}{k!(n-k)!}$ different combinations. In each of these combinations c_i, the scores of k relevant documents are denoted as $s_{i1}, s_{i2}, ..., s_{ik}$, and the scores of $n - k$ irrelevant documents are denoted as $s'_{i1}, s'_{i2}, ..., s'_{i(n-k)}$. With these probability values, we can answer a series of questions. For example, the probability that resultant list L includes at least k relevant documents is $\sum_{i=k}^{n} Pr(i)$.

The second question is: how many documents do we need to check if we want to obtain k $(k > 0)$ relevant documents? Obviously it is better to check those documents in the same order as they are ranked. The following estimation is based on such an assumption. In order to obtain k relevant documents, the expected number of documents we need to check is m, where $\sum_{i=1}^{m} s_i \geq k$ and $\sum_{i=1}^{m-1} s_i < k$.

The third question is: how many relevant documents are not in the resultant list? In order to answer this question, we need to estimate the relevance probabilities of those documents that are not in the resultant list. This can be estimated by using the relevance probability scores of the resultant list. For example, the logistic function [18] and the cubic function [109] have been found useful for such a purpose. See Sections 3.3.4 and 3.3.5 for details.

Example 2.4.1. A resultant list includes 12 documents. Their scores are $\{0.60, 0.56, 0.52, 0.48, 0.44, 0.40, 0.36, 0.32, 0.28, 0.24, 0.20, 0.16\}$.

A point estimate of the number of relevant documents in this list is

$$\sum_{i=1}^{12} s_i = 4.56 \approx 5$$

If we want to find 2 relevant documents, a point estimate of the number of documents we need to check is 4 since

$$\sum_{i=1}^{4} s_i = 2.16 > 2$$

and

$$\sum_{i=1}^{3} s_i = 1.68 < 2 \qquad \square$$

Another type of score can be provided is relevance score that indicates how relevant a document is. this explanation fits well when graded relevance judgment is applied.

If scores of relevance degree are provided, we can use the Euclidean distance as a measure to evaluate the effectiveness of any result. For a result $L = \{d_1, d_2, ..., d_n\}$, $S = \{s_1, s_2, ..., s_n\}$ and $H = \{h_1, h_2, ..., h_n\}$ are relevance scores provided by an information retrieval system and by human judge(s) for L, respectively. The effectiveness of S can be defined as the Euclidean distance between S and E

$$dist(S,H) = \sqrt{\sum_{i=1}^{n} (s_i - h_i)^2} \qquad (2.15)$$

Example 2.4.2. Assume that there are three documents in the result and their system-privided scores and human-judged scores are $S = \{1.0, 0.7, 0.2\}$ and $H = \{0.0, 0.4, 0.5\}$, respectively, then

$$dist(S,H) = \sqrt{(1.0 - 0.0)^2 + (0.7 - 0.4)^2 + (0.2 - 0.5)^2} = 1.09 \qquad \square$$

Example 2.4.3. Let us consider one run, $input.fub04DE$, which was submitted to the TREC 2004 robust track. 1000 documents were retrieved for each query. We estimate the observed curve by a logistic curve.

The logistic model uses the following function

$$Pr(t) = \frac{1}{1 + e^{-a - b*ln(t)}} \qquad (2.16)$$

to estimate curves (see Section 3.3.4 and 3.3.7 for more details). In Equation 2.16, t denotes a ranking position, $Pr(t)$ denotes the probability of the document at rank t being relevant to the query.

Table 2.1 Estimated number vs. real numbers of relevant documents at ranks 501-1000 for 8 randomly chosen runs in TREC 2004 ($error = |estimated - real|/real$)

Run	Estimated	Real	Error
apl04rsTDNfw	8.31	7.74	7.36%
apl04rsTDNw5	6.48	6.29	3.02%
humR04d4e5	7.31	6.09	20.03%
icl04pos2d	7.44	7.51	0.93%
JuruDes	7.19	6.53	10.11%
uogRobSWR10	9.24	7.24	27.62%
vrumtitle	8.73	7.25	20.41%
wdoqla1	8.03	7.33	9.55%

What we are doing is: first we use the top-500 documents to generate the estimated curve, then we use the same curve to estimate the number of relevant documents in the bottom-500 documents, and finally we compare the estimated numbers with real numbers of documents to see how accurate the estimation is.

We obtain the values for coefficients a and b in Equation 2.16. They are 1.6772 for a and -0.8682 for b. Therefore, we can estimate the relevance probability of a document at ranks 501-1000 using Equation 2.16 with two parameters a and b. As a step further, we can estimate the number of relevant documents for those documents ranking from 501 to 1000. The estimated number is 8.63 while the real number is 7.56 (a total of 1882 for 249 queries) for $input.fub04De$. Besides this run, we randomly chose 8 other runs and compared their estimated numbers with real numbers of relevant documents at ranks 500-1000, which are shown in Table 2.1. It seems that the estimation is quite accurate and the logistic function is a good function for this purpose.   □

It is possible to define variants of the Euclidean distance. Sometimes the users are just interested in a few top-ranked documents in the whole resultant list to see if they are useful or not. This is especially the case when the retrieved result comprises a large number of documents and it is difficult to read and judge all the documents in the result. Therefore, precision at given document cutoff values (say $m = 5$, 10, or 20) is a common user-oriented measure. Accordingly, we may also define the Euclidean distance of the m top-ranked documents as:

$$dist_{top\_m}(S_m, H_m) = \sqrt{\sum_{i=1}^{m}(s_i - h_i)^2} \tag{2.17}$$

Another variant of the Euclidean distance can also be defined. Suppose for the information need in question, $B_m = \{b_1, b_2, ..., b_m\}$ are the judged scores of m most relevant documents in the whole collection and $S_m = \{s_1, s_2, ..., s_m\}$ are judged scores of the m top-ranked documents in the resultant list. Then we may define the distance between S_m and H_m

$$dist_{best\_m}(S_m, B_m) = \sqrt{\sum_{best\_m}(s_i - b_i)^2} \tag{2.18}$$

This measure is analogous to recall in a sense that the top m documents in result L are compared with the m most relevant documents in the collection.

2.5 Uncertainty of Data Fusion Methods on Effectiveness Improvement

As observed in many experiments before, the uncertainty problem of data fusion methods is a norm. This is mainly caused by the metric used. If we use ranking-based metrics for retrieval results evaluation, then the problem of uncertainty is unavoidable. That means, in some cases, the performance of the fused result by a data fusion method can be better than the best component result; in some cases, the performance of the fused result by the same data fusion method can be worse than the worst component result.

For ranking-based data fusion methods such as the Condorcet voting and the weighted Condorcet voting, the problem of uncertainty is unavoidable since there is no score involved and we have to use ranking-based metrics to evaluate them.

It is possible for us to fuse results by a score-based data fusion method and evaluate them by a score-based metric, then the performance of the fused result is a determined problem. There is no uncertainty. For example, if we are going to fuse two component results by CombSum, then effectiveness of the fused result depends on both results. And it should be at least as good as the average of the two component results if the Euclidean distance is used. However, the fused result can go any direction if a ranking-based metric is used. see Section 6 for detailed discussion of score-based metrics for retrieval evaluation, See Section 7 for detailed discussion on ranking-based data fusion methods.

Chapter 3
Score Normalization

Score normalization is relevant to data fusion since very often those scores provided by component systems are not comparable or there is no scoring information at all. Therefore, score normalization is very often served as a preliminary step to data fusion. Score normalization methods can be divided into two categories: linear and non-linear. If no scores are provided, then it is possible to use some methods that can transform ranking into scores. In each case, we discuss several different methods.

Most data fusion methods proposed in information retrieval are score-based methods. That means, relevance scores are required for all retrieved documents in each component retrieval system. However, sometimes information retrieval systems only provide a ranked list of documents as the result to a query, and no scoring information is available. Sometimes information retrieval systems do provide scores for those retrieved documents, but the scores from different information retrieval systems may not be comparable since they are generated in different ways and can be different from scale to distribution. Therefore, in such a situation, we need to normalize those scores to make them comparable, before we are able to fuse them using a data fusion method. We discuss different ways of dealing with this issue.

In this chapter, we assume that binary relevance judgment is applied. Therefore, documents are divided into two categories: relevant and irrelevant. The only exception is Section 2.5.2, in which we shall discuss the score normalization issue when other types of relevance judgment, than binary relevance judgment, are applied.

Ideally, if there is a linear relation between score and probability of relevance, then it is a perfect condition for data fusion algorithms such as CombSum. More specifically, there are two types of comparability for n results L_i $(i = 1, 2, ..., n)$ retrieved from n information retrieval systems for the same information need:

- (**Comparability inside a result**) L is any one of the results, d_1 and d_2 are two randomly selected documents in L. d_1 obtains a normalized score of s_1, and d_2 obtains a normalized score of s_2, then the probability of d_1 being relevant to the information need equals to s_1/s_2 times of the probability of d_2 being relevant.
- (**Comparability across results**) for any two results L_1 and L_2 in n results, d_1 is a document in L_1 and d_2 is a document in L_2. If d_1's normalized score is equal

S. Wu: Data Fusion in Information Retrieval, ALO 13, pp. 19–42.
springerlink.com   © Springer-Verlag Berlin Heidelberg 2012

to d_2's normalized score, then these two documents have an equal probability of being relevant.

The above two types of comparability can be used to test how good a score normalization method is.

3.1 Linear Score Normalization Methods

Linear score normalization methods are straightforward. They normalize scores across different information retrieval systems into the same range or different but comparable ranges based on some rules. The linear score normalization method can be very successful if the component systems involved use linear functions to calculate scores.

3.1.1 The Zero-One Linear Method

Suppose there are m component results L_i, each of which is from an information retrieval system ir_i and contains n documents d_{ij} ($1 \leq i \leq m$) and ($1 \leq j \leq n$). r_{ij} is the raw score that d_{ij} obtains from ir_i. One straightforward way is to normalize scores of the documents in each L_i into the range of [0, 1] linearly. This can be done by using the following function

$$s_{ij} = \frac{r_{ij} - min\_r_i}{max\_r_i - min\_r_i} \qquad (3.1)$$

where min_r_i is the minimal score that appears in L_i, max_r_i is the maximal score that appears in L_i, r_{ij} is the raw score of document d_{ij}, and s_{ij} is the normalized score for document d_{ij}.

One basic requirement for using this method properly is: for those top-ranked documents in the resultant list, their probability of relevance should be close to 1; for those bottom-ranked documents in the resultant list, their probability of relevance should be close to 0. Therefore, in order to achieve this, we require that:

- Every component system is at least reasonably good so as to be able to identify some of the relevant documents at top ranks.
- If there are quite a large number of relevant documents in the whole collection, then the length of the resultant list should be long enough then the bottom-ranked documents are very likely to be irrelevant.

If the two conditions listed above are satisfied, then we are quite sure that top-ranked documents are very likely to be relevant and bottom-ranked documents are very likely to be irrelevant. Thus the scores normalized by the zero-one method can be used for CombSum to achieve good fusion results. Otherwise, it is better for us to use a different score normalization method (such as the fitting method, see next subsection) or a different data fusion method (such as the linear combination method) to improve effectiveness.

The zero-one linear score normalization method has been used in many experiments, e.g., in [48] and others. The above zero-one linear method can be improved in two different ways, which will be discussed in the next two subsections.

3.1.2 The Fitting Method

In many situations, [0, 1] may not be the best interval for score normalization. We consider the situation of TREC, in which each information retrieval system provides up to 1000 documents in any resultant list. The number of relevant documents varies across different queries, from as less as 1 or 2 to over one hundred. Even if all 1000 documents are used for fusion, top-ranked documents are not always relevant and bottom-ranked documents are not always irrelevant. Therefore, to normalize all the scores in any resultant list into a range $[a, b]$ $(0 < a < b < 1)$ should be a better option than [0, 1]. Moreover, if we use fewer documents (for example, top 100, or even top 10) for fusion, then the zero-one score normalization method becomes more questionable, since the bottom-ranked documents are far from irrelevant in such a situation.

Generally speaking, if used properly, the fitting method can lead to better normalized scores than the zero-one score normalization method does. However, it is obvious that some training is needed to decide the values of a and b properly. Linear regression can be used to decide the best possible values of a and b when some training data are available. For a group of results L_i $(1 \leq i \leq n)$, suppose that max_s_i and min_s_i are maximum and minimum scores in L_i $(1 \leq i \leq n)$, respectively. After dividing (min_s_i, max_s_i) into m equal-width intervals, we can put documents into m buckets, each of which is corresponding to an interval. For example, suppose that the maximum and the minimum scores are 100 and 0 in one result, and we divide the whole range into 20 intervals. Then all the documents whose scores are between 100 and 95 (including 100 but not including 95) are put into bucket 1, all the documents whose scores are between 95 and 90 (including 95 but not including 90) are put into bucket 2, ..., and so on, and all the documents whose scores are between 5 and 0 (including both 5 and 0) are put into bucket 20. Note that each resultant list may have a different score range from the others, but we can always divide the whole range into the same number of intervals. All the documents in each bucket needs to be examined to see what percentage of them are relevant. Then a linear regression can be carried out to obtain the best values of a and b. Another option is to observe some top-ranked and bottom-ranked documents and estimate the probabilities of them being relevant and use them as the values of a and b. This can be done more easily and the estimation can still be quite accurate.

The fitting method was investigated in [117] and used in a few experiments including [26] and others.

3.1.3 Normalizing Scores over a Group of Queries

Another way of improving the zero-one method is to define the global normalization score over a group of queries from a given information retrieval system. Suppose there are a group of queries $Q = (q_1, q_2,..., q_p)$, every information retrieval system ir_i returns a ranked list L_i^k for each of the queries q_k ($1 \leq k \leq p$). Assume that $min\_r_i^k$ is the minimal raw score that appears in L_i^k, $max\_r_i^k$ is the maximal raw score that appears in L_i^k. We define min_r_i and max_r_i as

$$max\_r_i = max\{max\_r_i^1, max\_r_i^2, ..., max\_r_i^p\} \tag{3.2}$$

$$min\_r_i = min\{min\_r_i^1, min\_r_i^2, ..., min\_r_i^p\} \tag{3.3}$$

max is the function that takes the maximal value, while min is the function that takes the minimal value from a list of values. Then any raw score can be normalized by

$$s_{ij}^k = \frac{r_{ij}^k - min\_r_i}{max\_r_i - min\_r_i} \tag{3.4}$$

3.1.4 Z-Scores

In statistics, a standard score, known as Z-score, indicates how many standard deviations a datum is above or below the mean. For a group of raw scores $\{r_1, r_2, ..., r_n\}$ from a ranked list of documents L, let us assume that μ is the simple mean, or $\mu = \frac{1}{n} \sum_{i=1}^n r_i$, and σ is the standard deviation, or $\sigma = \sqrt{\frac{1}{n} \sum_i^n (r_i - \mu)^2}$, then the normalized score s_i for any r_i is defined as

$$s_i = \frac{r_i - \mu}{\sigma} \tag{3.5}$$

The range of Z-scores is $(-\infty, \infty)$. Z-scores may be used for CombSum and the linear combination method, but they are not very suitable for CombMNZ. In statistics, Z-scores are used to normalize observed frequency distribution of a random variable, which is very different from the situation of document scores here. Different information retrieval systems may use very different schemas (e.g., estimated probability of the document's relevance to the information need, odds ratio of the estimated probability, or the natural logarithm of the odds ratio of the estimated probability) to score documents. Therefore, some results may be close in performance but their score distribution may be very different from each other. In such a situation Z-scores are unlikely to distinguish them adequately. Previous experiments [60, 45, 79] show modest success of this method with the TREC data.

Negative scores are not good for some data fusion methods such as CombMNZ. A variant of the Z-score method is to add a positive number (e.g., 1 or 2) to scores that are already normalized by the Z-score method. They are referred to as Z1 or Z2 according to the number (1 or 2) added.

3.1.5 Sum-to-One

Suppose there are a group of raw scores $\{r_1, r_2, ..., r_n\}$, from a ranked list of documents, the sum-to-one method normalizes raw scores using

$$s_i = \frac{r_i}{\sum_{i=1}^{n} r_i} \tag{3.6}$$

The sum-to-one method is similar to the normalization method used in the Demster-Shafer theory of evidence, in which the masses of all members are added up to a total of 1. This normalization method was investigated in [60].

Next let us analyse sum-to-one. Suppose we have n component systems, the normalized scores of result L_i range from 0 to a_i ($0 < a_i < 1$ for $i = 1, 2,..., n$), and we use CombSum for the fusion process. As a matter of fact, we may use zero-one to normalize these results, and then use the linear combination method with $\{a_i\}$ to be the weights of these results $\{L_i\}$ for $(i = 1, 2, ..., n)$ to achieve exactly the same effect. It suggests that we should be able to estimate the performance of a result by considering some statistics of the scores of its documents, if sum-to-one can work well. However, it is not clear if we can set up a relation between these two factors.

We carry out an experiment to investigate the correlation between the performance of a component result and the range of its normalized scores using sum-to-one. As we analysed before, there must be some correlation between them for sum-to-one to outperform zero-one. We use four groups of results, which are results submitted to TRECs 7 (ad hoc track), 8 (ad hoc track), 9 (web track), and 2001 (web track). The conditions for the selection are:

- 1000 documents are provided for every query;
- the average performance (measured by average precision) of the result over 50 queries is over 0.15.

The first condition is applied since some runs include fewer documents, and removing them makes the processing easier and experimental results more reliable. By applying the second condition, we remove those poor results in which we are not interested. The threshold of 0.15 has been chosen arbitrarily.

As a matter of fact, in TRECs 9 and 2001, three-graded relevance judgment is used. That means, all documents are divided into three categories: highly relevant, modestly relevant, and irrelevant. However, here we stick to binary relevance judgment. We do not distinguish modestly relevant documents from highly relevant documents and treat them equally as relevant documents.

For a selected group of results submitted to TRECs 7, 8, 9, and 2001, we normalize all of them using sum-to-one and record the maximal normalized scores. Then for the whole collection, we calculate the correlation coefficient k (Kendall's tau-b) of these two variables, which are averaged over 50 queries. The range of k is [-1, 1], with 1 for identical ranks and -1 for opposite ranks. The results are shown in Table 3.1.

Table 3.1 Kendall's tau-b correlation coefficients between the performance of a component result and the range of its normalized scores using sum-to-one

Group	Correlation Coefficient (Significance)
TREC 7	0.067(0.448)
TREC 8	0.318(0.003)
TREC 9	-0.102(0.383)
TREC 2001	0.080(0.532)

In Table 3.1, we can see that modest positive correlation exists in TREC 8, and no correlation can be observed in the three other groups. It suggests that there is no strong positive correlation between the two factors.

3.1.6   Comparison of Four Methods

In summary, all linear score normalization methods are very straightforward. The major characteristic of the linear score normalization methods is that only the range, but not the distribution of the original scores is changed. Therefore, how successful the linear score normalization is largely depends on the relationship between relevance and distribution of original scores, although there are certain difference among different linear score normalization methods. In this subsection, we make a comparison of the four methods discussed before: the zero-one method, the fitting method, sum-to-one, and the Z-score method.

For the experiment to examine how good these score normalization methods are, the procedure is as follows: first, for a result, we normalize the scores of its documents using a given score normalization method. Next we divide the whole range of scores into a certain number of intervals. Then for every interval, we count the number of relevant documents (*num*) and the total score (*t_score*) obtained by all the documents in that interval. In this way the ratio of *num* and *t_score* can be calculated for every interval. We compare the ratios of all internals so as to evaluate how good the score normalization method is.

We are not focused on any particular information retrieval system here. A more effective way to understand the scoring mechanism of a particular information retrieval system is to investigate the scoring algorithm used in that system, not just analyse some query results produced by it. Rather, we try to find out some collective behaviour of a relatively large number of information retrieval systems in an open environment such as TREC. In such a situation, to analyse a group of query results becomes a sensible solution. There are a few reasons for this: firstly, if the number of systems is large, it is a tedious task to understand the implementation mechanisms of all the systems involved; secondly, it is not possible if the technique used in some of the systems is not publicly available; thirdly, the collective behaviour of a group of systems may not be clear even we know each of them well.

Four group of results are used in the experiment. They are subsets of results submitted to TRECs 7 (ad hoc track), 8 (ad hoc track), 9 (web track), and 2001 (web track). Two conditions are applied to those runs submitted: 1000 documents are provided for each query and the average performance over 50 queries is 0.15 or above (measured by average precision).

We normalize all the results using the fitting method $[0.06\text{-}0.6]$[1], zero-one, sum-to-one and Z1 (a variant of the Z-score method, which adds 1 to normalized Z scores) over 50 queries. Then for all normalized results with a particular method, we divide them into 20 groups. Each group is corresponding to a interval and the scores of all the documents in that group are located in the same interval.

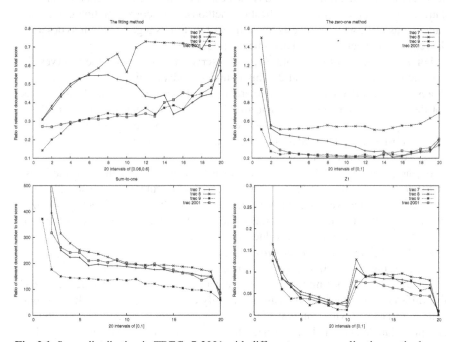

Fig. 3.1 Score distribution in TRECs 7-2001 with different score normalization methods

For the fitting method, we divide $[0.06, 0.6]$ into 20 equal intervals $[0.06, 0.087)$, $[0.087, 0.114),..., [0.573, 0.6]$. For zero-one, we divide $[0,1]$ into 20 equal intervals

---

[1] The values chosen for a and b ($a = 0.06$ and $b = 0.6$) are reasonable, but not optimal.

Table 3.2 Average relative differences of *num/t_score* for four groups of results (the figures in parentheses exclude data in Interval 1)

Group	Fitting[0.06-0.6]	zero-one	Sum	Z1
TREC 7	0.1550	0.3115(0.2202)	0.8974(0.1791)	1.8999(0.7105)
TREC 8	0.1733	0.1729(0.0522)	1.7028(0.2380)	1.8998(0.7135)
TREC 9	0.1426	0.1478(0.0870)	0.2347(0.1745)	1.8994(0.7057)
TREC 2001	0.2073	0.2842(0.1520)	0.6176(0.1818)	1.8996(0.4317)

[0, 0.05), [0.05, 0.1),..., [0.95, 1]. For sum-to-one, we divide [0, 1] into 20 intervals [0, 0.00005), [0.00005, 0.0001), ..., [0.00095, 1], since all normalized scores in sum-to-one are very small and most of them are smaller than 0.001. For Z1, 20 intervals used are (-∞, 0.1), [0.1, 0.2),..., [1.9, +∞). For each group, we calculate its ratio of *num* and *t_score*. All results are shown in Figure 3.1. The ideal situation is that the curve is parallel to the horizontal axis, which means a linear relation between score and document's relevance and CombSum is a perfect method for data fusion. It demonstrates that the fitting method is the best since all the curves generated are the flattest. For all other score normalization methods, their corresponding curves are quite flat as well, which suggests that a linear function is a good option to describe the relation between score and document's relevance.

For zero-one, sum-to-one and Z1, we observe that very often their curves fly into the sky in Intervals 1 and 2. This is understandable since all the normalized scores with zero-one and sum-to-one are approaching 0 in Intervals 1 and 2. It is not the case for their raw scores and there are a few relevant documents in these two intervals. The curves are deformed because the corresponding scores are under-valued considerably. The situation is even worse for Z1 since negative and positive scores coexist in Interval 1.

Furthermore, let us take a more careful look at these normalization methods' ability to provide a linear relation between score and document's relevance. For the same group of data, we obtained 20 values (*ratio=num/t_score*) for each normalization method. We average these 20 ratio values and calculate the relative difference of all values to the average ratio values by using

$$relative\_difference = |ratio - ave\_ratio|/ave\_ratio$$

then we average these 20 relative differences. The results are shown in Table 3.2.

In Table 3.2, The fitting method and the zero-one method lead to the least relative difference in almost all the cases. The only exception happens to the zero-one method in TREC 7 when Interval 1 is not considered. Since the curves of [0.06-0.6] are quite flat, Interval 1 does not need to be excluded as in other methods. Z1 is always the worst whether we include Interval 1 or not.

3.2   Nonlinear Score Normalization Methods

The linear score normalization methods look a little naive, since different information retrieval systems may use very different formulae to generate scores for the
documents retrieved. If we want not only to change the range, but also the distribution of the raw scores, then nonlinear functions are required. Some such functions
have been proposed.

3.2.1   The Normal-Exponential Mixture Model

One approach is the mixture model proposed by Manmatha, Rath, and Feng [55].
In their mixture model, the set of non-relevant documents is modelled by an ex-
ponential distribution and the set of relevant documents is modelled by a normal
distribution.Then a standard approach, expectation maximization(EM), is used to
find the mixing parameters and the parameters of the component densities. They experimented with two information retrieval systems INQUERY and SMART and two
groups of data sets used in TREC 3 and TREC 4. The results show that the method
is quite successful. However, there is one theoretical flaw about this mixture model:
the estimated probability of relevance as a function of the score is non-monotonic.

3.2.2   The CDF-Based Method

It is a two-step process. First two functions need to be defined. One is the Cumulative
Density Function (CDF) for any information retrieval system ir_i involved. This can
be obtained by using ir_i working on a group of training queries. The other function
is the Optimal Score Distribution (OSD). This function is vaguely defined as the
score distribution of an ideal scoring function that matches the ranking by actual
relevance. How to obtain an appropriate OSD is critical to the method and there may
be different ways to obtain the OSD. In [30], the proposed method is to compute the
average distribution of several good retrieval systems, in which the zero-one linear
method is used to normalize raw scores. Note that in this method, we need to define a
CDF function for every information retrieval system involved, while all component
systems may share the same OSD function.

Given the two functions F_τ for CDF and F for OSD, we are able to normalize
any raw score, as the second step. For any raw scores r_i, the normalized score is $s_i =
F^{-1}(F_\tau(r_i))$. The process of normalization is illustrated in Figure 3.2 with two raw
score examples r_1 and r_2. The normalized scores of them are s_1 and s_2, respectively.

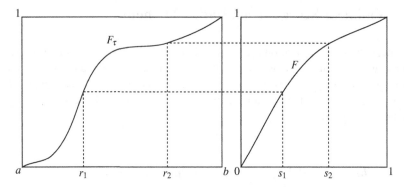

Fig. 3.2 The process of normalizing raw scores of the CDF-based method

3.2.3 Bayesian Inference-Based Method

Given the ranked lists of documents returned by n retrieval systems, let $t_i(d)$ be the rank assigned to document d by retrieval system ir_i. This constitutes the evidence of relevance provided to the merging strategy concerning document d. For a given document, let $P_{rel} = Pr[rel|t_1, t_2, ..., t_n]$ and $P_{irr} = Pr[irr|t_1, t_2, ..., t_n]$ be the respective probabilities that the given document is relevant and irrelevant given the rank evidence $t_1, t_2, \ldots, t_n$. We compute the log odds of relevance: $logO[rel] = log(P_{rel}/P_{irr})$ and rank documents according to this measure. According to [41], we have

$$logO[rel|t_1, t_2, ..., t_n] = \sum_{i=1}^{n} logO[rel|t_i] - (n-1)logO[rel]$$

This formula can be proven by induction as follows [59]. The base case is trivial. Now if we assume that the formula holds for $n-1$, we can show that it holds for n.

$$O[rel|t_1...t_n] = \frac{Pr[rel|t_1...t_n]}{Pr[irr|t_1...t_n]}$$

$$= \frac{Pr[t_1...t_n|rel]Pr[rel]}{Pr[t_1...t_n|irr]Pr[irr]}$$

$$= \frac{Pr[t_1...t_{n-1}|rel]Pr[t_n|rel]Pr[rel]}{Pr[t_1...t_{n-1}|irr]Pr[t_n|irr]Pr[irr]}$$

(independence is assumed in the above)

$$= \frac{Pr[rel|t_1...t_{n-1}]Pr[rel|t_n]Pr[irr]}{Pr[irr|t_1...t_{n-1}]Pr[irr|t_n]Pr[rel]}$$

$$= \frac{O[rel|t_1...t_{n-1}]O[rel|t_n]}{O[rel]}$$

Therefore,

$$logO[rel|t_1...t_n] = logO[rel|t_1...t_{n-1}] + logO[rel|t_n] - logO[rel]$$

Applying our inductive hypothesis yields the desired result.

Dropping terms that are the same for all documents gives a formula for calculating the log odds of relevance for ranking:

$$s(d) = \sum_{i=1}^{n} logO[rel|t_i]$$

Note that either rankings or scores can be used in the above formula. For better estimation, training is required for parameters setting.

*   *   *

In the above we have discussed both linear and non-linear score normalization methods. Compared with the linear combination methods, non-linear combination methods are more complicated but have the potential to achieve better results. However, due to the diversity of score generating methods used in different information retrieval systems, special treatment for each information retrieval system is very likely required if we wish to achieve proper normalization effect.

3.3 Transforming Ranking Information into Scores

It happens quite often that information retrieval systems only provide ranked lists of documents without any scores. In such a situation, we may use ranking-based methods such as the Condorcet voting to fuse results. Thus scores are not needed. Another option is to convert ranking information into scores. A few models have been proposed in this category.

3.3.1 Observing from the Data

Without any assumption of the function that exists between rank and probability of relevance, we may simply observe that from some training data [51]. This can be done for any individual information retrieval system or a group of information retrieval systems together. To make the observation reliable, a significant number of documents at each rank need to be provided and evaluated. One straightforward way is to calculate the posterior probability of relevance of documents at every rank. That is, for a set of queries $Q = \{q_1, q_2,..., q_n\}$, the probability of relevance for documents at rank t is

$$Pr(t) = \frac{\sum_{i=1}^{n} f(t,i)}{n} \qquad (3.7)$$

where $f(t,i)$ is a function. If a document at rank t is relevant to query q_i, then $f(t,i)$ = 1; otherwise $f(t,i) = 0$.

When the number of documents evaluated are not large enough, it is very likely that the probability curve does not look normal. The probability at some ranks may be as lower as 0, the probability at a rank may be lower than the probability at successive ranks, etc. We can always treat (e.g., smooth, interpolate and so on) the probability curve in one way or another to make them look better.

One possible treatment is to use a sliding window of the fixed size [50, 51]. After obtaining the initial probability at each rank $Pr(t)$ by Equation 3.7, the probability can be smoothed by

$$Pr'(t) = \frac{\sum_{i=t-u}^{t+u} Pr(t)}{2u+1} \qquad (3.8)$$

Here $2u + 1$ is the size of the sliding window. Apart from ranking position t, u ranking positions on both left and right sides are considered as well. At some very top ranks and very bottom ranks, fewer neighbouring positions are used if not all required exist.

Another possible solution is that sliding windows may be defined to be of variable size.

3.3.2   The Borda Count Model

In the Borda count, documents are given scores based on their ranks. For a ranked list of n documents, the first one is given n points, the second one is given $n - 1$ points, and so on,..., and the last one is given 1 point. Then score-based methods such as ComSum, CombMNZ, or the linear combination method, can be used for fusion.

It may be arguable why such a linear weighting assignment function, rather than a quadratic function, a cubic function, or even a logarithmic function, is used. For some applications such as political voting, there is no way to justify which function is the best, which is not. However, for the data fusion problem in information retrieval, we are in a better position to perceive which function is good and which is not. This is because in information retrieval, all documents are divided into different categories, for example, two categories, if binary relevance judgment is used. The problem of score assignment from ranking becomes to find out the relationship between probability of relevance and ranks. It is possible to find some good functions by theoretical analysis and/or empirical investigation. Many functions have been tried and some of them seem to be good candidates. In the following let us discuss more options.

3.3.3 The Reciprocal Function of Rank

For a group of ranked documents $<d_1, d_2,..., d_n>$, whose ranks are 1, 2,..., n, respectively. The scores are normalized by $s_i = 1/i$ for $(1 \leq i \leq n)$.

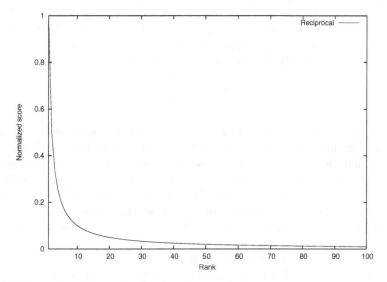

Fig. 3.3 The reciprocal normalization function of rank

Figure 3.3 shows the curve of the reciprocal function. The function decreases with rank very rapidly.

The reciprocal function is proposed by Lillis et. al. in [50]. After normalization, a performance-level weighting (measured by average precision) is applied to each of the information retrieval systems involved, and then the linear combination method is used. The fusion method as such is referred to as MAPFuse [50].

3.3.4 The Logistic Model

The logistic model is investigated by Calve and Savoy [18] in the context of distributed information retrieval. Then, some experiments [79, 80] have also been conducted to investigate the effectiveness of the method for data fusion.

$$s(t) = \frac{e^{a+b*t}}{1+e^{a+b*t}} = \frac{1}{1+e^{-a-b*t}} \qquad (3.9)$$

In Equation 3.9, a and b are two parameters. In [18], they use $ln(t)$ to replace t in the above equation. Usually, the logistic function is a S-shape curve. However, its shape varies with different a and b values. According to [18], the rationale of using the logarithm of rank instead of the rank itself is based on the following consideration: "retrieval schemes generally display documents in a descending order (from more relevant to less relevant), therefore the probability of relevance changes systematically with the rank order (or the serial order). However, using this ordinal variable without any transformation assumes regular differences between retrieved document positions which is not realistic. A difference of 10 in ranks seems to be more significant between the ranks 20 and 30 than between 990 and 1000. These last ranks contain such a small number of relevant documents that it might be appropriate to ignore the difference between 990 and 1000[2]." Therefore, the following equation

$$s(t) = \frac{e^{a+b*ln(t)}}{1+e^{a+b*ln(t)}} = \frac{1}{1+e^{-a-b*ln(t)}} \qquad (3.10)$$

is used. The function in Equation 3.10 is the composition of a logarithm function and a logistic function, which we refer to as the modified logistic function. The curve of the modified logistic function is somewhat different from that of a pure logistic function. Figure 3.4 shows a example of two curves, in which $a = 1$ and $b = -0.75$. They are typical values for our purpose. Both curves decrease monotonously when rank increases. However, compared with the logistic function, the modified logistic function decreases at a much slower pace. This is understandable because $ln(t)$ is used instead of t. When $t = 1$, $ln(t) = 0$. The highest point of the curve is totally decided by a. When t is larger, b's effect on the curve becomes more significant.

Looking at Figure 3.4, the logistic curve decreases very quickly to very close to 0 when rank is 10 or above. That is why the logistic model does not, while the modified logistic model works well for the score normalization problem.

3.3.5   The Cubic Model

The cubic model was investigated in [109]. Just as in the modified logistic model, we use $ln(t)$ to replace t. The (modified) cubic model uses the following function

$$s(t) = a + b * ln(t) + c * ln(t)^2 + d * ln(t)^3 \qquad (3.11)$$

---

[2] However, this may not be the key point. See below and Section 2.3.6 for more comments.

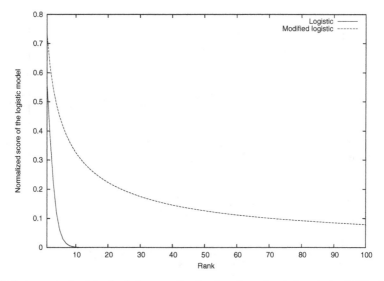

Fig. 3.4 An example of the logistic normalization function of rank ($a = 1$, $b =-0.75$)

to estimate the relation between rank and probability of relevance. Figure 3.5 shows a curve of the cubic model, in which $a = 0.4137$, $b = -0.0699$, $c = -0.0049$, $d = 0.0009$. These are typical values [3].

In the experiments conducted in [109], the results show that the modified cubic model and the modified logistic model are almost equally effective. This should not be very surprising. If we compare the curves in Figures 3.4 and 3.5, both of them are very much alike each other.

However, there is two possible problems with the cubic function. One problem is the function is not always monotonically decreasing with t or $ln(t)$. For example, the curve shown in Figure 3.5 has a positive $d = 0.0009$. When t is large enough, $d * ln(t)^3$ increases faster than other terms and the function will monotonically increase with t. The second problem is, there is no guarantee that the normalized scores are always positive. Sometimes we may obtain some negative scores for certain ranks. Usually, these two problems does not affect the performance of data fusion results very much, if only a certain number of top-ranked documents are used.

3.3.6 The Informetric Distribution

In the last couple of decades, a family of distributions have been found very useful for describing a wide range of phenomena in different disciplines. Many people,

---

[3] The values are obtained by regression (curve estimation) with a group of runs submitted to the web track in TREC 9. More details can be seen in [109].

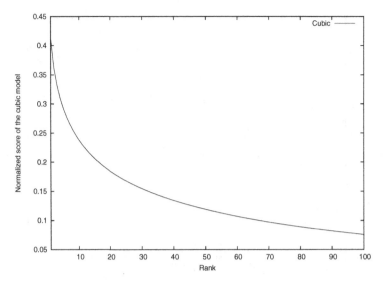

Fig. 3.5 An example of the cubic normalization function of rank ($a = 0.4137$, $b = -0.0699$, $c = -0.0049$, $d = 0.0009$)

including Zipf, Bradford, Lotka, and others, made significant contribution to them. In information science, there have been a lot of studies about using those distributions to describe various things. For example, Zipf's law [131] can be used to describe the distribution of words in a document [5].

According to [13, 14], although a family of regularities have been found to describe a wide range of phenomena both within and outside of the information sciences, these distributions are just variants of a single distribution, which is

$$s(t) = \frac{c}{t^b} \tag{3.12}$$

where t is the rank, c is a constant, and b is a parameter ranging normally between 1 and 2. It is referred to as the informetric distribution [13, 14].

It was considered that Equation 3.12 may be used as an alternative way of describing relevance distributions [91]. In the following, let us look at the relation between those that have been investigated for fusion and the informetric distribution mathematically.

In Equation 3.12, if we let $c = 1$ and $b = 1$, then we obtain the reciprocal function of rank. Therefore, the reciprocal function of rank is obviously a simplified version of Equation 3.12.

Next we consider the modified logistic model, which was discussed in Section 2.3.4. Equation 3.10 can be rewritten as

$$s(t) = \frac{1}{1 + e^{-a - b' * \ln(t)}} = \frac{1}{1 + e^{-a} * t^{-b'}} \tag{3.13}$$

If we remove the additive 1 from the denominator and let $e^a = c$ and $-b' = b$, then we have

$$s(t) \approx \frac{1}{e^{-a} * t^{-b'}} = \frac{e^a}{t^{-b'}} = \frac{c}{t^b} \tag{3.14}$$

The right side of Equation 3.14 is exactly the same as that of Equation 3.12. Therefore, we can see that the difference between Equations 3.12 and 3.13 is small, when t is large ($1 << e^{-a} * t^{-b'}$). But they are quite different when t is small.

Equation 3.12 can be rewritten as

$$ln(s(t)) = ln(c) - b * ln(t) \tag{3.15}$$

Equation 3.15 can be used to find parameters $ln(c)$ and b by linear regression with some training data which provide observed relevance distribution. See next subsection for more related discussion.

It seems that there is no relation between the modified cubic model and the informetric distribution. However, we should bear in mind that any functions can be estimated by a polynomial, especially with large degree terms. Replacing t by $ln(t)$ makes the curve smoother and easier for the cubic model to fit.

In the rest of this book, the modified logistic model and the modified cubic model are referred to as the logistic model and the cubic model, respectively. Since for converting ranking information into scores, we do not use the (original) logistic model and the (original) cubic model anyway and no confusion should arise.

3.3.7 Empirical Investigation

In this subsection we investigate and evaluate all the methods discussed in this section by experiments. Two groups of results are used. They are selected subsets from all the runs submitted to the ad hoc track in TREC 8 and the web track in TREC 9.

The two groups are quite different in both size and performance. For the TREC 8 group, there are 101 runs. The average performance (measured by AP) of all of them is 0.4849; for the TREC 9 group, there are only 44 runs. The average performance (also measured by AP) is only 0.2671. We intentionally choose those two quite different groups so as to let the investigation have a wider coverage of different situations.

Since each run only provides lists of documents for 50 queries, it is not enough for us to obtain reliable observations if we investigate those runs individually. Therefore, in a group, we put all the runs together to generate a super run. Thus we are able to investigate the collective property of them.

For the super-run of a group of results, we calculate the observed relevance distribution, or the relevance score for each of the ranking positions. The process is like this: for the TREC 9 case, there are 2200 documents (44 runs * 50 queries) at each position. We evaluate how many of them are relevant. If at a position t, there are 220 relevant documents, then the relevance score at this position is 220/2200 = 0.1. We work out relevance scores for all 1000 ranking positions as such. The process for the TREC 8 group is the same. Table 3.3 (a) and (b) show the first 5 records of two groups (TREC 8 and TREC 9), respectively. Apart from rank and score, ln(rank) (the logarithm of rank with respective to base e) and ln(score) are also included because they need to be used as well.

Table 3.3 Relevance distributions of two groups of results (TREC 8 and TREC 9)

rank	score	ln(rank)	ln(score)
1	0.3800	0.0000	-0.9676
2	0.3400	0.6931	-1.0788
3	0.3222	1.0986	-1.1326
4	0.2968	1.3863	-1.2147
5	0.2695	1.6094	-1.3112
..	...	...	...

(a) The TREC 8 group

rank	score	ln(rank)	ln(score)
1	0.5778	0.0000	-0.5485
2	0.5394	0.6931	-0.6173
3	0.5112	1.0986	-0.6710
4	0.4857	1.3863	-0.7222
5	0.4512	1.6094	-0.7958
..	...	...	...

(b) The TREC 9 group

The two score distributions are shown in Figure 3.6. It can be clearly seen the score difference between the two groups of TREC 8 and TREC 9.

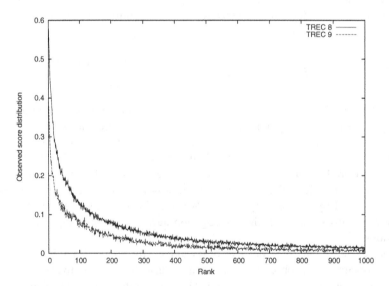

Fig. 3.6 The observed score distributions of two groups of selected runs (TRECs 8 and 9)

Next let us find the best possible values for those parameters needed. Five models are considered. They are Reciprocal, Borda, Informetric, Logistic, and Cubic. For the reciprocal and Borda models, estimated scores can be generated directly. For example, for the reciprocal model, 1/1=1.0000, 1/2=0.5000, 1/3=0.3333, 1/4=0.2500,

1/5=0.2000,..., are estimated scores for documents at ranking positions 1, 2, 3, 4, 5,..., etc. For the three others, more work is required.

For the informetric, logistic, and cubic models, we try to find the best suitable parameters for them. This can be done by using the observed relevance distribution to carry out regression analysis.

For the cubic model, we use *score* as the dependent variable, and ln(*rank*) as the independent variable to run the curve estimation (regression)[4]. The values of a, b, c, and d we obtain are shown in Table 3.4 (a).

For the informetric model, we use ln(score) as the dependent variable and ln(rank) as the independent variable to run linear regression (see Equation 3.15). The values of parameters b and c (see Equation 3.12) are shown in Table 3.4 (b).

For the logistic model, a binary logistic regression is carried out to decide the values of a and b (see Equation 3.10). Thus a different data file than Table 3.3 is required. The data record is this data file includes two columns. The first is *rank*, which gives the ranking position of the document, the second is *relevance*, which indicates if the document is relevant or not. If it is a relevant document, then 1 is assigned; otherwise, 0 is assigned. For every ranked document in each component result, we evaluate it and generate a record for it. The values of parameters of b and c (see Equation 3.10) are shown in Table 3.4 (c).

Table 3.4 Parameters obtained for three different models (cubic, informetric, and logistic)

Group	a	b	c	d
TREC 8	0.608	-0.082	-0.013	0.002
TREC 9	0.384	-0.064	-0.004	0.001

(a) The cubic model

Group	b	c
TREC 8	0.811	4.568
TREC 9	0.885	3.669

(b) The informetric model

Group	a	b
TREC 8	1.516	-0.786
TREC 9	0.937	-0.781

(c) The logistic model

From Table 3.4, we can see that the parameters we obtain for two groups of results are consistent. For example, in the cubic model, the values of a d are positive and the values of b and c are negative in both groups; in the logistic model, the values of b are negative and also very close to each other.

With the parameters, we are able to generate estimated relevance scores at all ranks using any of the three models. Next we compare the estimated relevance scores with the observed ones to see how accurate the estimated relevance scores are. We use linear regression again for this purpose. This time observed score is used as the independent variable and estimated one as the dependent variable. The statistics for the linear regression analysis are shown in Tables 3.5 and 3.6. They are R^2, F, and the significance level.

From Tables 3.5 and 3.6 we can see that in all the cases, curve estimation does a good job since the significance (p-value) is at the .000 level, or smaller than 0.0005. But this measure is not useful for us to distinguish one model from another. The two

---

[4] SPSS is used in this experiment. Its web site is located at
http://www-01.ibm.com/software/uk/analytics/spss/

Table 3.5 Statistics of the linear regression analysis (observed score used as independent variable and estimated one as dependent variable, the TREC 8 data set)

Regression model	R^2	F	Significance level
Reciprocal	0.398	660.477	.000
Borda	0.501	1004.008	.000
Informetric	0.559	1262.930	.000
Logistic	0.957	22284.930	.000
Cubic	0.975	38256.819	.000

Table 3.6 Statistics of the linear regression analysis (observed score used as independent variable and estimated one as dependent variable, the TREC 9 data set)

Regression model	R^2	F	Significance level
Reciprocal	0.422	729.285	.000
Borda	0.494	974.110	.000
Informetric	0.512	1049.907	.000
Logistic	0.900	8995.719	.000
Cubic	0.901	9036.440	.000

other measures R^2 and F are very helpful. When the two variables have strong linear relationship to each other, we expect high R^2 and F values. Thus we can see both the cubic model and the logistic model are the most accurate models, the Borda model and the reciprocal model are the least accurate ones, while the informetric model is in the middle but only slightly better than the Borda model and the reciprocal model.

It is not surprising that Reciprocal and Borda are not as good as the other three models since they do not need any training. However, if we compare the reciprocal model and the Borda model, it is a little surprising for us to perceive that the reciprocal model is not as good as the Borda model. Let us look at them more closely. Figure 3.7 shows the estimated distributions and the observed ones for the TREC 8 group. The situation is very similar for the TREC 9 group.

In Figure 3.7, the observed curve is in the middle of the curve for the Borda model and the one for the reciprocal model. It tells us that the Borda model always over-estimates the relevance scores while the reciprocal model always under-estimates the relevance scores. It seems that the curve of the reciprocal model is closer to the observed one than that of the Borda model, but that does not help. At first the reciprocal curve drops very quickly when ranking position increases, then comes to a turning corner, after that it drops very slowly. Thus the reciprocal curve is very close to both horizontal and vertical axes.

The informetric model is worth another look. The definitions of the informetric model and the logistic model are very similar, but their performance are very different. This phenomenon is interesting. Figure 3.8 [5] shows the estimated distributions of the informetric model and the logistic model and also the observed one for the TREC 8 group. The situation is very similar for the TREC 9 group again.

---

[5] In order to see the difference clearly, only the top 100 ranks are displayed.

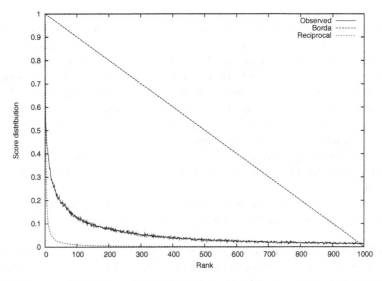

Fig. 3.7 The score distributions of the TREC 8 data set by using two different models (Borda, Reciprocal)

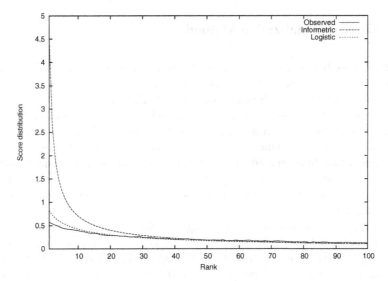

Fig. 3.8 The score distributions of the TREC 8 data set estimated by using the informetric model and the logistic model

In Figure 3.8, we can see that when the ranking position becomes 50 or more, both informetric curve and logistic curve fit the observed curve very well. We are sure it keeps as such until the last ranking position (1000), even Figure 3.8 does not show them all (refer to Equation 3.13 and Equation 3.14). However, at the very beginning, the informetric model over-estimates the relevance scores considerably, while the logistic model over-estimates the relevance scores slightly. Such a difference makes the logistic model very good while the informetric model not very good. Note that for data fusion, the accuracy of scores for those top-ranked documents is more important than other parts of a result.

The logistic model performs as good as the cubic model in TREC 9, but it is not as good as the cubic model in TREC 8. In the latter case, the difference between them is not large.

It is notable that for all those models, their performances in both groups are consistent. It also suggests that the results are very reliable about the comparison of them.

Finally, let us compare the same model in the two groups. Every model performs better in TREC 8 than in TREC 9 except the reciprocal model. A hypothesis is: apart from the reciprocal model, those models involved are more accurate when the results are more effective. We leave this as a research issue for the future.

3.4 Mixed Normalization Methods

In the above we have discussed quite a few different score normalization methods. It is possible for us to use a linear combination of different normalization methods, instead of any individual method, to achieve even better results.

Basically, many of those normalization methods can be used together. For example, one may use a combination of two methods, one of which is a linear score normalization method, and the other is to convert ranks to scores. Another example is to combine two different methods which convert ranks to scores. If one finds that method A is good at the top-ranked part and B is good at the bottom-ranked part, then one can set up a separating point, and use each of them in a range at which the method is good.

Sholouhi proposed a mixed method, SegFuse, for score normalization in [82]. Scores are generated using a mixture of normalized raw scores (s_{n-rs}, raw scores are provided by the information retrieval system) and ranking-related relevance scores (s_p, from ranking-related posterior probabilities). For s_{n-rs}, the 0-1 linear score normalization method is used to normalize raw scores. For obtaining s_p, SegFuse needs a group of training queries and a document collection. SegFuse divides the whole lists of documents into segments of different size ($size_k = (10 * 2^{k-1} - 5)$). Thus, $size_1 = 5$, $size_2 = 15$, $size_3 = 35$, and so on. In each segment, documents at different ranks are assigned equal relevance scores. The final score of a document is given by $s_p(1 + s_{n-rs})$. For the fusion process, SegFuse just uses the same method as CombSum does.

3.5 Related Issues

In this section we discuss two related issues: weights assignment vs. score normalization, and relevance judgments other than binary relevance judgment.

3.5.1 Weights Assignment and Normalization

There are connections between CombSum, the linear combination method, score normalization, and weights assignment. First of all, if we compare the definitions of CombSum (Equation 1.1), and the linear combination method (Equation 1.3), we can see that CombSum is a special form of the linear combination method in which all weights are equal to 1. Secondly, for both CombSum and the linear combination method, score normalization is needed to make scores comparable across different information retrieval systems. Thirdly, weights assignment is only needed for the linear combination method, but not CombSum. Fourthly, it may be possible for a single process to combine score normalization and weights assignment in some cases. In the following, let us discuss some of the implications.

As we will see later (Chapters 4-6) in this book, for a group of information retrieval systems, the weight for every system is decided by its performance and/or similarity between it and others. This needs to be taken into consideration when we try to combine score normalization and weights assignment for the lienar combination methods.

In Sections 3.1-3.4, we have discussed a range of score normalization methods. Some methods are more sophisticated than others. Though under the title of "score normalization", some sophisticated score normalization methods do have a component of weighting assignment for different information retrieval systems. This is because in some cases, system performance is taken into consideration, but similarity between systems has been touched by none of them.

Some of the methods such as the zero-one linear method, global scores, Z-scores, sum-to-one, the Borda model, and the reciprocal model do not consider system performance. Some other methods such as the fitting method, the normal-exponential mixture model, the CEF-based method, the logistic model, the cubic model, the informetric model, mixed normalization methods, or even simply observing from the data, are taking system performance into consideration. Therefore, we can expect that in general, all the methods in this latter category can be more effective than the methods in the former category.

3.5.2 Other Relevance Judgments

So far all the score normalization methods have been discussed under the assumption that binary relevance judgement is used. When other types of relevance judgment are used, the usability of those score normalization methods discussed in this chapter need to be re-examined. In the following, let us consider this problem if graded relevance judgment is used. For graded relevance judgment, See Section 2.3 for detailed discussion.

For $n + 1$ graded relevance judgment, there are $n + 1$ grades of relevance. Grade n is the most relevant, which is followed by $n - 1$, $n - 2$,..., 1, while grade 0 is irrelevant. A usual treatment for n graded relevance judgment is: if a document is relevant at grade i ($1 \leq i \leq n$), then we assign a score of i/n to i. After that, all the linear score normalization methods discussed in Section 3.1 can be used. However, all those nonlinear score normalization methods (see Section 3.2) cannot be used anymore. It is not clear how those methods can be adapted in the new environment. Finally, for those methods that transform ranking information into scores, most of them can still be used, but binary logistic model is an exception.

One related issue is the performance of those score normalization methods when graded relevance judgment is used. Next let us see some preliminary findings about this. Two groups of TREC results (9 and 2001 web track) are used. Both of them were involved in the comparison experiment in Section 3.1.6. As a matter of fact, both of them use the 3-graded relevance judgment. That is to say, any documents are divided into three categories: highly relevant, modestly relevant, and irrelevant. But in Section 3.1.6, both modestly relevant and highly relevant documents are treated equally as relevant documents. Thus 3-graded relevance judgment becomes binary relevance judgment. Here we use the original 3-relevance judgment to investigate the score normalization issue.

The cubic model is used to fit the curves of rank-relevance in the condition of binary and 3-graded relevance judgments. The statistics of the curve estimation are shown in Table 3.7.

Table 3.7 Statistics of the rank-relevance curve estimation of the cubic model

Relevance judgment	TREC 9		TREC 2001	
	R^2	F	R^2	F
Binary	0.991	35383.012	0.990	31803.081
3 graded	0.992	42728.444	0.988	27378.482

In Table 3.7 we can see that the curve estimation is almost equally good for both binary relevance judgment and 3 graded relevance judgment. In all the cases, the curve estimation is very accurate (in both cases R^2 is very close to 1).

Chapter 4
Observations and Analyses

Due to the uncertainty involved, it is difficult to answer questions such as why and how data fusion can improve retrieval performance, which data fusion method is better, or in what condition the data fusion methods can improve retrieval performance. However, in the last two decades, some effort has been taken to try to find some sort of answer to these questions. It is understandable that statistical analysis plays a very important role. In this chapter, we are going to discuss some progress already made in this regard.

4.1 Prior Observations and Points of View

In later 1980s, researchers in the information retrieval community began to investigate the data fusion issue. Saracevic and Kantor [78] investigated multiple results from the same information retrieval system, but independently-generated query representations, and found that different query representations led to different groups of documents. They also found that a document was more likely to be relevant if it appeared in more results.

Turtle and Croft [96] performed similar experiments using an inference network, and found that combining two different query representations (probabilistic and Boolean versions of queries) led to increased retrieval performance over any single query representation. They originally thought that at least part of the performance improvements arose because the two query types were retrieving different relevant documents, so that the combined set contained more relevant documents than retrieved by the separate queries. This is not, however, the case. The documents retrieved by the Boolean queries are a subset of those retrieved by the corresponding probabilistic queries. Foltz and Dumais [31] found similar improvements by combining different retrieval models.

Belkin et. al. [11, 12] carried out data fusion experiments with a large 2GB document collection (for the time) used in TREC 1. They found performance improvements consistent with the prior work of the time. However, they observed "that

S. Wu: Data Fusion in Information Retrieval, ALO 13, pp. 43–71.
springerlink.com   © Springer-Verlag Berlin Heidelberg 2012

different representations of the same query, or of the documents in the database, or different retrieval techniques for the same query, retrieve different sets of documents (both relevant and irrelevant)".

Fox and his colleagues [32, 33] investigated a group of fusion methods including CombSum and CombMNZ. They found that CombSum and CombMNZ outperformed the others. They also found that using different retrieval techniques, query representations, or document representations often led to result sets with surprisingly little overlap, which was consistent with several research efforts prior to their own [46, 78].

At this point, people began to believe that data fusion could be an effective technique for improving retrieval performance considerably. Probably the only negative result in the 1990s was reported by Thompson [94], who used the linear combination method to fuse results submitted to TREC 1. Thompson found that the combined results, weighted by performance level, performed no better than a combination using uniform weights (CombSum). However, this does not necessarily mean that the data fusion techniques are not effective in general.

Lee [48] did experiments with six results submitted to TREC 3. They are $brkly6$, $eth001$, $nyuir1$, $pircs1$, $vtc5s2$, and $westp1$. All the selected runs are close in performance, and none of the two runs are very similar since they are submitted by different research groups[1]. He found that in all the combinations, CombSum and CombMNZ outperformed the best component system involved and CombMNZ outperformed CombSum slightly. Lee proposed a hypothesis that as long as the component systems used for fusion had greater relevant overlap than irrelevant overlap, performance improvement would be achievable. He also demonstrated that the hypothesis held for the data set he used.

Vogt and Cottrell [98, 99] investigated the linear combination problem of two systems with all the runs submitted to the TREC 5 ad hoc track. For each query and each pair of systems, a single parameter α ($\sin(\alpha)$ is used as the weight for the first result and $\cos(\alpha)$ is used as the weight for the second result) was chosen by optimising average precision using golden section search [71], and the best α was used to generate the fused result.

Three kinds of effects were mentioned. they are:

- **The skimming effect** happens when different retrieval systems retrieve different relevant documents, so that a fusion method that takes the top-ranked documents from each of the retrieval systems will push non-relevant documents down in the rankings.
- **The chorus effect** occurs when several retrieval systems suggest that a document is relevant to a query, this tends to be stronger evidence for relevance than a single system doing so.

---

[1] In TREC, each research group is allowed to submit multiple runs to the same track. Usually those multiple runs are generated from the same information retrieval system, but with different parameter settings, different query formats, etc. Those runs submitted by the same group are much more similar than those submitted by different research groups.

- **The dark horse effect** means that a retrieval system may produce unusually accurate (or inaccurate) estimates of relevance for at least some documents, relative to the other retrieval systems.

However, they noted that when choosing how to combine the results from different information retrieval systems, the dark horse effect is at odds with the chorus effect; on the other hand, a large chorus effect cuts into the possible gain from the skimming effect. These phenomena suggest that only a sophisticated fusion method may be able to predict when these effects will occur and take advantage of them. Such methods would almost certainly need to make use of training data in the form of user feedback in order to fine tune their performance.

Vogt and Cottrell also proposed their hypothesis for the fusion of two component results. That is, linear combination is warranted when

- at least one system exhibits good performance;
- both systems return similar sets of relevant documents;
- both systems return dissimilar sets of irrelevant documents.

We can clearly see the similarity between Lee's and Vogt and Cottrell's hypotheses.

Beitzel et al. [10] conducted some experiments to compare the performances of CombMNZ using several different groups of systems. They observed no improvement when fusing results from three different retrieval models in the same information retrieval system, while the merged result was better than the best system when choosing the top three systems submitted to TREC 6, 7, 8, 9 and 2001. In all the cases involved, relevant overlap was greater than non-relevant overlap. Therefore, they argued that greater overlap of relevant documents than of non-relevant documents, which was proposed by Lee, was not a very good indicator for fusion improvement.

Soerri [89, 90] also investigated the data fusion issue using 5 TREC data sets (TRECs 3, 7, 8, 2003, and 2004). He mentioned two effects: the authority and ranking effects. He claimed that both of them played a key role in data fusion. The authority effect refers to the fact that the potential relevance of a document increases exponentially as the number of systems retrieving it increases and the ranking effect refers to the phenomena that documents higher up in ranked lists and found by more systems are more likely to be relevant.

Nassar and Kannan [63] summarized a few factors that affect the performance of data fusion results. All the factors are classified into three categories:

- design of data fusion methods: mainly includes the skimming effect, the chorus effect, and the dark horse effect;
- properties of individual component results: overlap rate among the results, number of component results, average performance of the results, standard deviation of the performance of the results, and performance of the best component result;
- features used as input to the data fusion methods: normalized score, rank, and both normalized score and rank.

Although there have been quite a few different hypotheses, they are not very helpful for us to understand the data fusion problem. It is also not clear how we can use them to improve fusion performance.

4.2 Performance Prediction of CombSum and CombMNZ

How to predict the performance of data fusion algorithms is a difficult task due to the uncertainty of the problem. In this section, we address this issue by statistical analysis.

Vogt and Cottrell [98, 99] analysed the performance of the linear combination algorithm using linear regression. In their experiments, two systems were always used for fusion, which is the simplest situation. 14 variables were used in the analysis: two different performance measures (one of them was average precision and the other was a statistical measure of rank correlation between the system and the relevance judgement) of each system, the number of relevant documents returned by one system but not the other divided by the total number of relevant documents returned by that system, the similarity of two results' rankings and others. The performance analysis and prediction for the fused result was very accurate ($R^2 = 0.94$). However, for the prediction of performance improvement of the fused result over the best component system, their analysis model was not useful ($R^2 = 0.06$).

Ng and Kantor [64] focused on predicting if the performance of CombSum is better than all component systems or not. They used several different statistical techniques: linear analysis, multiple linear regression, logistic regression and a nonparametric training and testing method which they called the bin-ranking method. They also used two systems for each fusion, as Vogt and Cottrell did in [98, 99]. Two variables were used: performance ratio of two systems, and a measure of the dissimilarity between two systems. They found that the two variables were informative to predict if the fused result was better than both component systems and the detection rate of their approach was about 70% to 75%. Also they used multiple linear regression to predict the performance improvement of the fused result over the better one among two results. A R^2 of 0.204 was observed.

Wu and McClean investigated the performance prediction issue of data fusion methods including CombSum and CombMNZ in [123]. In this section we mainly take materials from [123], with some necessary updates.

The overall approach is to run several fusion algorithms with a large number of combinations of results from actual IR systems, and to identify the variables, via multiple regression, that affect the performance of data fusion algorithms. With a few independent variables and one dependent variable, a multiple linear regression attempts to fit a linear model to data. In such a way, we can find if there is a relation between the dependent variable and all the independent variables or how strong the relation is. Again, TREC data is very suitable for our purpose. We choose three groups of information retrieval results, the first is a subset of 42 results submitted to the TREC 6 ad hoc track, the second is a subset of 58 results submitted to the TREC 2001 web track, and the third is a subset of 77 results submitted to the TREC 2004 robust track[2]. All selected runs include 1000 documents for each query and have a performance of above 0.15 by average precision.

From a group of results, we randomly choose a certain number (3, 4, 5, 6, 7, 8, 9, or 10) of results, then we randomly choose 10,000 combinations for every given number n ($3 \leq n \leq 10$). Three fusion methods, CombSum (see Equation 1.1), CombMNZ (see Equation 1.2) and Round-robin, are used in the experiment. Round-robin works like this: it takes one document from every result in turn, deleting any document if it has occurred in the fused result before. Although Round-robin is not a usual data fusion method, we include it in the investigation in order to use it as a baseline and make a comparison between it and common data fusion methods such as CombSum and CombMNZ. The zero-one score normalization method is used for all the results before the fusion process.

We consider several aspects: the average performance of all component systems (*ave_p*), the standard deviation of the performance of all component systems (*dev*), the number of results (*num*), and the correlation among component results (*o_rate*). For performance evaluation, we use average precision, since it is a single value measure and convenient for us to use.

We calculate the average performance (measured by average precision) of every result over a certain number of queries (50 for TREC 6 and TREC 2001, 249 for TREC 2004); and for each combination, we calculate the standard deviation of their average performance. How to decide the strength of correlation among two or more component results is a question that needs to be considered carefully. Note here we are not concerned about the difference/similarity of information retrieval systems which are used to retrieve documents, but only the final document results, though there is strong relation between the result we obtain from an information retrieval system for a given query and the information retrieval system itself (including many aspects such as retrieval models, query formations, parameter settings, and so on). We may have several different ways of calculating the correlation coefficient of two results over the same group of queries (e.g., Spearman correlation coefficient, Kendall's tau measure). However, we need to calculate $n(n-1)/2$ correlation coefficients for n systems. Moreover, it is difficult to calculate the correlation among more than 2 results. Instead of using correlation coefficients, we therefore calculate the overlap rate among a group of results as

$$o\_rate = \frac{D_{all} - D_{unique}}{D_{all}} \tag{4.1}$$

where D_{all} is the number of documents in all the results, and D_{unique} is the number of documents which only occur in any one of the results but not the others. We use this *o_rate* to describe the correlation strength among a group of results.

We set several different objectives (performance of the fused result, performance improvement rate of the fused result to the average of all component results, and

---

[2] 250 queries were used in the TREC 2004 robust track. However, there is one query (number 672) whose relevant document set is not included in the official relevance judgements file "qrels.robust2004.txt". Therefore, we use 249 queries.

performance improvement of the fused result to the best component result) as dependent variables in the multiple regression analysis to observe the effect on them of those defined variables.

All the variables used in this section with their meanings are listed in Table 4.1.

Table 4.1 Variables and their meanings used in Section 4.2

num	number of results for fusion
o_rate	overlap rate among all component results
ave_p	average performance of all component results
dev	standard deviation of performance of all component results
best	performance of the best component result
fused	performance of the fused result
imp	a Boolean variable indicating if the fused result is better than the best of two results
first	performance of the first component result
second	performance of the second component result
diss	dissimilarity between two results
ratio	performance ratio of two results

4.2.1 Data Fusion Performance

Let us see performance analysis first. Table 4.2, 4.3, and 4.4 present the performances of CombSum, CombMNZ, and Round-robin for TREC 6, TREC 2001, and TREC 2004 respectively. The standardized coefficients of the resulting regression equation can be interpreted as indicating how much each variable contributes to the overall estimate of the dependant variable. Thus, a positive coefficient indicates that the corresponding variable should be maximized in order to maximize the performance. Conversely, a negative coefficient indicates that the variable should be minimized in order to maximize the performance. The actual coefficients of the regression equation is standardised based on the distribution of the individual independent variables, so that their magnitude can be compared. R^2 measures how well we can predict the performance knowing only the four independent variables in the model. For example, if the value of R^2 is 0.65, it means that the four variables explain 65% of the variation in the performance of the data fusion method. From Tables 4.2 - 4.4, we can also observe that significance for all independent variables is listed at the .000 level, which means that the p value is less than 0.0005 and all independent variables are statistically highly significant with a probability of over 99.95% (1 - 0.0005 = 99.95%). All coefficients are consistent across three different groups of data sets, including their sign and ranking of absolute values.

Comparing CombSum and CombMNZ, we find that they are very similar in many ways:

- All corresponding variables take similar values in both methods.
- Their R^2 values are close.

Table 4.2 Effect of 4 variables on the performance of data fusion methods (TREC 6)

Variable	Standardized coefficients		
	CombSum	CombMNZ	Round-robin
num	0.445	0.484	0.200
o_rate	-0.283	-0.327	-0.082
ave_p	0.576	0.613	0.800
dev	0.309	0.258	0.175
	$R^2 = 0.848$	$R^2 = 0.852$	$R^2 = 0.924$

Significance: .000 for all variables in all three methods.

Table 4.3 Effect of 4 variables on the performance of data fusion methods (TREC 2001)

Variable	Standardized coefficients		
	CombSum	CombMNZ	Round-robin
num	0.625	0.630	0.221
o_rate	-0.365	-0.402	-0.140
ave_p	0.646	0.669	0.668
dev	0.177	0.172	0.283
	$R^2 = 0.816$	$R^2 = 0.807$	$R^2 = 0.837$

Significance: .000 for all variables in all three methods.

Table 4.4 Effect of 4 variables on the performance of data fusion methods (TREC 2004)

Variable	Standardized coefficients		
	CombSum	CombMNZ	Round-robin
num	0.727	0.714	0.188
o_rate	-0.392	-0.412	-0.155
ave_p	0.931	0.929	0.915
dev	0.269	0.217	0.121
	$R^2 = 0.778$	$R^2 = 0.791$	$R^2 = 0.894$

Significance: .000 for all variables in all three methods.

- The Pearson correlation coefficients for the results of CombMNZ and CombSum are 0.933 (TREC 6), 0.990 (TREC 2001), and 0.991 (TREC 2004) which indicate that these two methods are strongly correlated to each other.
- The average performance (measured by average precision) of 80,000 combinations over 50 queries are 0.3061 and 0.3051 for CombSum and CombMNZ, respectively in TREC 6; and they are 0.2551 and 0.2555 in TREC 2001. In TREC 2004, the figures over 249 queries are 0.3461 and 0.3431 for CombSum and CombMNZ, respectively.

Though all the variables are highly significant, their effects on the performance of the fused results are not the same. According to the absolute values of coefficients, we can rank the 4 variables in descending order according to their significance. For all 3 methods, the average performance of all results (ave_p) is always the most significant variable. However, in Round-robin, ave_p is in the dominating position since its coefficient value is much larger than the others. This situation does not happen in either CombSum or CombMNZ. This is understandable because Round-robin fuses all component results in such a way that the order of a document is totally determined by its original position in one of the component results. In all three methods, o_rate takes a negative value. This indicates that overlapping is harmful to the performance of data fusion in all cases. However, the effect of overlapping on these three methods is not the same. CombMNZ is the most sensitive one; Round-robin is the least sensitive one; while CombSum is in the middle. This is because CombMNZ heavily uses the "multiple evidence principle", which arranges the documents retrieved by multiple results in high priority, while Round-robin does not do this at all. When overlap rate is high, which means the results involved are not very different, to use methods such as CombMNZ cannot boost much the performance of the fused result.

The above multiple linear regression analysis assumes that all the relations are linear, which may not be appropriate for all the variables. Therefore, we tried some variations. One variation is, instead of using only dev, we use both dev and dev^2, instead of using num, we use $\ln(num)$, instead of using only o_rate, we use both o_rate and o_rate^2, also we use both ave_p and $\text{sqrt}(ave\_p)$ to replace ave_p. See Table 4.5 for detailed information about the regression analysis result of the TREC 6 document collection. The results of the two other collections are not presented since they are analogous to the one presented.

Table 4.5 Regression of data fusion performance handling nonlinear relationships (TREC 6)

Variable	Standardized coefficients		
	CombSum	CombMNZ	Round-robin
$ln(num)$	0.636	0.679	0.249
o_rate	0.953	0.831	0.195
o_rate^2	-1.336	-1.268	-0.334
ave_p	2.024	1.783	0.597
$sqrt(ave\_p)$	-1.418	0.647	0.204
dev	-0.456	-0.463	0.307
dev^2	0.738	0.693	-0.147
	$R^2 = 0.914$	$R^2 = 0.916$	$R^2 = 0.930$

Significance: .000 for all the variables in all three methods.

For all three collections, the changes made lead to considerable improvement for both CombSum ($R^2 = 0.914$ for TREC 6, 0.872 for TREC 2001, and 0.860) and combMNZ ($R^2 = 0.916$ and 0.863 and 0.871), but only very slight improvement ($R^2 = 0.930$ and 0.848 and 910) for Round-robin.

It demonstrates that the linear model is not the best model for this problem. For those independent variables, we can add polynomial or other types of terms to make the linear model more accurate. Furthermore, it is very likely that the best model for this problem is intrinsically nonlinear [35]. We can see this point more clearly in Section 6.

4.2.2 Improvement over Average Performance

Another issue is which variables may lead to performance improvement for data fusion methods. We continue to use multiple regression to investigate this. All the variables used in the above analysis are kept the same; however, we change the dependent variable from fusion performance to performance improvement (percentage of performance improvement of data fusion over average performance of all component results).

Tables 4.6-4.8 present the results for all three fusion methods. All four variables are statistically highly significant with only one exception. Comparing Tables 4.6, 4.7 and 4.8 with Tables 4.2, 4.3, and 4.4, the orders of significance of these variables are very different. Instead of average performance (ave_p), num becomes the most significant variable, while ave_p becomes the least significant one in all the cases. In both CombSum and combMNZ, all these variables are ranked in the same order. Both o_rate and ave_p take negative values, which means that overlap rate and average performance should be kept at the minimal level in order to obtain maximal performance improvement. Conversely, standard deviation of average performance (dev) has a fairly large positive coefficient, and it should be boosted for performance improvement.

Table 4.6 Effect of several variables on performance improvement of data fusion methods over average performance (TREC 6)

Variable	Standardized coefficients		
	CombSum	CombMNZ	Round-robin
num	0.827	0.916	0.567
o_rate	-0.519	-0.612	-0.237
ave_p	-0.112	-0.094	-0.079
dev	0.500	0.419	0.471
	$R^2 = 0.565$	$R^2 = 0.553$	$R^2 = 0.404$

Significance: .000 for all variables in all three methods.

Table 4.7 Effect of several variables on performance improvement of data fusion methods over average performance (TREC 2001)

Variable	Standardized Coefficients		
	CombSum	CombMNZ	Round-robin
num	1.015	1.055	0.434
o_rate	-0.595	-0.636	-0.276
ave_p	-0.004	-0.032	-0.022
dev	0.260	0.246	0.532
	$R^2 = 0.534$	$R^2 = 0.537$	$R^2 = 0.395$

Significance: .000 for all variables (except ave_p in CombSum: 0.282) in all three methods.

Table 4.8 Effect of several variables on performance improvement of data fusion methods over average performance (TREC 2004)

Variable	Standardized coefficients		
	CombSum	CombMNZ	Round-robin
num	0.824	0.849	0.474
o_rate	-0.469	-0.468	-0.397
ave_p	-0.246	-0.265	-0.483
dev	0.316	0.680	0.313
	$R^2 = 0.693$	$R^2 = 0.680$	$R^2 = 0.467$

Significance: .000 for all variables in all three methods.

Let us consider an example. Suppose we have two groups of results, with each group including 5 results, and the overlap rates of these two groups are the same. Five results in the first group have an average precision of 0.2, 0.2, 0.3, 0.4, 0.4, respectively, and the average precision is 0.3 for every result in the second group. According to regression analysis, the first group is more likely to obtain better fusion result than the second group even though their average performance is the same in both groups. We explain this phenomenon like this: if some results are better than some others, then these good results are more likely to share some common opinion, and their common opinion will dominate the whole group; while those poor results share less common opinion, and their effect on fusion is limited. On the other hand, if all the results are close in performance, then no one result or several results can dominate the whole group, and less improvement can be made by data fusion.

Further improvement of the model is possible as in Section 4.2.1. When we use dev, dev^2, num, $ln(num)$, o_rate, o_rate^2, ave_p, and sqrt(ave_p), improvement can be observed for all methods in all situations. The R^2 values become 0.755 (TREC 6) and 0.688 (TREC 2001) and 0.804 (TREC 2004) for CombSum, 0.745 (TREC 6) and 0.683 (TREC 2001) and 0.797 (TREC 2004) for CombMNZ, and 0.461 (TREC 6) and 0.443 (TREC 2001) and 0.506 (TREC 2004) for Round-robin. Since all values of R^2 are larger here than those with linear variables, the predictions here are more accurate.

We also observe that fused result is almost always better than the average of component results. The opposite situation rarely happens. Out of 240,000 combinations, 8 times it occurs for CombSum, 6 times for CombMNZ, and 25 times for Round robin. This demonstrates that data fusion methods are effective on improving the performance of the fused result over the average of component results.

4.2.3   Performance Improvement over Best Performance

We now use the performance improvement of data fusion over the best component result as the dependent variable to run the multiple regression analysis. Tables 4.9-4.11 show the results with five linear variables: num, o_rate, ave_p, dev, and $best$.

We observe that increasing the number of component results and increasing average performance of all component results are helpful, while higher overlap rate among results, diversified performances of component results, and especially lofty best results are very harmful for data fusion methods to outperform the best component result.

As in Sections 4.2.1 and 4.2.2, we can also increase the values of R^2 in these methods by introducing non-linear variables. The R^2 values become 0.820 (TREC 6) and 0.864 (TREC 2001) and 0.811 (TREC 2004) for CombSum, 0.839 (TREC 6) and 0.863 (TREC 2001) and 0.835 (TREC 2004) for CombMNZ, and 0.840 (TREC 6) and 0.905 (TREC 2001) and 0.922 (TREC 2004) for Round-robin. Since all values of R^2 are bigger here than those in Section 4.2.2, the predictions here are more accurate than that in Section 4.2.2.

Table 4.9 Effect of several variables on the performance improvement of data fusion methods over best system (TREC 6)

Variable	Standardized coefficients		
	CombSum	CombMNZ	Round-robin
num	0.747	0.498	0.067
o_rate	-0.454	-0.487	-0.157
ave_p	0.517	0.519	0.543
dev	-0.143	-0.184	-0.265
best	-1.083	-1.071	-1.044
	$R^2 = 0.654$	$R^2 = 0.694$	$R^2 = 0.889$

Significance: .000 for all variables in all three methods.

Table 4.10 Effect of several variables on the performance improvement of data fusion methods over best system (TREC 2001)

Variable	Standardized coefficients		
	CombSum	CombMNZ	Round-robin
num	0.543	0.571	-0.070
o_rate	-0.395	-0.422	-0.170
ave_p	0.470	0.438	0.417
dev	-0.114	-0.116	-0.210
best	-1.169	-1.150	-0.938
	$R^2 = 0.788$	$R^2 = 0.791$	$R^2 = 0.827$

Significance: .000 for all variables in all three methods.

Out of 80,000 combinations, 35,206 (44.0%) of the fused results using Comb-Sum outperform the best component result, and 35,143 (43.9%) of the fused results using CombMNZ outperform the best component result in TREC 6. In TREC 2001, these two figures are 57,959 (72.4%) and 58,514 (73.1%). In TREC 2004, they are 68160 (85.2%) and 64321 (80.4%). Therefore, there is an approximately 67% chance that we can observe that CombSum and CombMNZ are better than the best

Table 4.11 Effect of several variables on the performance improvement of data fusion methods over best system (TREC 2004)

Variable	Standardized coefficients		
	CombSum	CombMNZ	Round-robin
num	0.853	0.820	0.153
o_rate	-0.488	-0.453	-0.167
ave_p	1.088	1.075	0.711
dev	0.317	0.264	0.289
best	-1.208	-1.201	-0.766
	$R^2 = 0.695$	$R^2 = 0.727$	$R^2 = 0.738$

Significance: .000 for all variables in all three methods.

component result. For Round-robin, the figures are 11,514(14.4%), 21149(26.4%), and 15203(19.0%), in TREC 6, TREC 2001, and TREC 2004, respectively. We also notice that the figures in TREC 6 are lower than that in TREC 2001 and TREC 2004. This is because a few component systems in TREC 6 are much better than the others, while the performances of all component results in TREC 2001 and TREC 2004 are close.

4.2.4 Performance Prediction

We divide 50 queries in TREC 6 into two parts, the first part includes the first 25 queries and the second part the second 25 queries. The first half is used for training, and the second half is used for prediction. We calculate the average performance (measured by average precision) for every combination and every fused result, and then compare them with real values. The relative errors for CombSum, CombMNZ, and Round-robin are 0.0310, 0.0358, and 0.0278, respectively.

Next we use all 80,000 combinations in TREC 6 for training, and then use the coefficients obtained to predict the performance of 80,000 combinations of TREC 2001. The relative errors for CombSum, CombMNZ, and Round-robin are 0.0575, 0.0570, and 0.0350, respectively. Considering that the two groups of systems, document collections, and queries are totally different, this suggests that the analytical result is still useful even when we apply it in a very different situation from that used in training.

Discriminate analysis is discussed in [64] and aims to predict if the fused result is better than the best component result. Our above analysis is applicable for the same purpose. For every combination, we calculate the average performance of the fused result *real_p*, and estimate the average performance of that *es_p* according to the multiple regression analysis, then we compare them with the average performance of the best result *best* to see how many times the judgement is correct by checking if $((es\_p > best)$ AND $(real\_p > best))$ OR $((es\_p < best)$ AND $(real\_p < best))$

holds. For CombSum, the detection rates are 90.0% (TREC 6) and 93.2% (TREC 2001); for CombMNZ, the detection rates are 90.6% (TREC 6) and 92.8% (TREC 2001). Our result is better than that in Ng and Kantor's work ([64]): 70% for testing runs and 75% for training runs.

If we don't have to make judgements for all the cases, then we can increase the correct detection rate by neglecting those cases which are on the margin of profit/loss for data fusion. Table 4.12 shows the detection rates of the prediction in various conditions for TREC 6 and TREC 2001. We check $((es\_p > (1+k)*best)$ AND $(real\_p > best))$ OR $((es\_p < (1+k)*best)$ AND $(real\_p < best))$ holds for how many combinations with different k (k = 0, 0,01,..., 0.10). Generally speaking, the prediction is more accurate when the condition is more restrictive.

Table 4.12 Detection rate in two different situations (TREC 6 & TREC 2001)

Condition	TREC 6		TREC 2001	
(k)	CombSum	CombMNZ	CombSum	CombMNZ
0.00	90.0%	90.5%	93.2%	92.8%
0.01	92.0%	93.2%	95.2%	94.8%
0.02	94.2%	95.2%	96.7%	96.4%
0.03	96.0%	96.7%	97.8%	97.6%
0.04	97.5%	97.8%	98.6%	98.5%
0.05	98.4%	98.6%	99,1%	99.0%
0.06	99.0%	99.1%	99.4%	99.4%
0.07	99.4%	99.4%	99.7%	99.7%
0.08	99.7%	99.7%	99.8%	99.8%
0.09	99.8%	99.7%	99.9%	99.9%
0.10	99.9%	99.9%	99.9%	99.9%

4.2.5 The Predictive Ability of Variables

Although the same multiple regression technique has been used in Ng and Kantor's work [64], we use different variables. This is why we are able to achieve much more accurate prediction. Therefore, it is interesting to conduct an experiment to compare the predictive ability of those variables used in their model and/or our model. In Ng and Kantor's work, they used two variables to predict the performance of the fused result with CombSum: (a) a list-based measure of result dissimilarity, and (b) a pair-wise measure of the similarity of performance of the two systems. The result dissimilarity of two systems is calculated as follows: for the same query, assume we obtain the same number (e.g., 1000) of retrieved documents from both systems. We merge these two results to obtain a larger group of documents (with n documents). For every possible combination of any two documents in this large group, we compare their respective rankings in both results. If the rankings are the same, a score of 0 is given; if the rankings are opposite, a score of 1 is given; if the situation is uncertain, a score of 0.5 is given, then we sum up all scores and divide it by $n(n-1)/2$, which is the maximal possible score for the two results. In this way we obtain a normalised score between 0 and 1 for any pair of results.

We used 42 systems in TREC 6 for the experiment. All possible pairs (861) of them are used for data fusion with CombSum, CombMNZ and Round-robin. We analyse these results using the multiple regression method with the same dependent but different independent variables. In such a way, we can decide the predictive ability of different variables. The experimental results are shown in Table 4.13.

Table 4.13 Predictive ability of different variables (TREC 6)

Number	Method	Dependent variable	Independent variables	R^2	Pair Comparison
1	CombSum	$fused$	$o\_rate, first, second$	0.932	2.31%
2	CombSum	$fused$	$diss, first, second$	0.911	
3	CombSum	$fused$	o_rate, ave_p, dev	0.915	1.67%
4	CombSum	$fused$	$diss, ave\_p, dev$	0.900	
5	CombSum	imp	$o\_rate, ratio$	0.409	10.84%
6	CombSum	imp	$diss, ratio$	0.369	
7	CombSum	imp	o_rate, ave_p, dev	0.399	11.76%
8	CombSum	imp	$diss, ave\_p, dev$	0.357	
9	CombMNZ	$fused$	$o\_rate, first, second$	0.927	3.11%
10	CombMNZ	$fused$	$diss, first, second$	0.899	
11	CombMNZ	$fused$	o_rate, ave_p, dev	0.909	2.60%
12	CombMNZ	$fused$	$diss, ave\_p, dev$	0.886	
13	CombMNZ	imp	$o\_rate, ratio$	0.403	13.52%
14	CombMNZ	imp	$diss, ratio$	0.355	
15	CombMNZ	imp	o_rate, ave_p, dev	0.401	14.57%
16	CombMNZ	imp	$diss, ave\_p, dev$	0.350	
17	Round-robin	$fused$	$o\_rate, first, second$	0.984	1.03%
18	Round-robin	$fused$	$diss, first, second$	0.974	
19	Round-robin	$fused$	o_rate, ave_p, dev	0.979	0.82%
20	Round-robin	$fused$	$diss, ave\_p, dev$	0.971	
21	Round-robin	imp	$o\_rate, ratio$	0.432	9.37%
22	Round-robin	imp	$diss, ratio$	0.395	
23	Round-robin	imp	o_rate, ave_p, dev	0.460	10.31%
24	Round-robin	imp	$diss, ave\_p, dev$	0.417	

Note: $fused$ denotes the performance (average precision) of the fused result, imp is a Boolean variable indicating if the fused result is better than the best of the two results, o_rate denotes overlap rate between two results, $first$ denotes the average precision of the first result, $second$ denotes the average precision of the second result, $diss$ denotes the dissimilarity measure between the two results, ave_p denotes the average performance of the two results, dev denotes the standard deviation of $first$ and $second$, $ratio$ denotes the ratio of performance of two results (the better one divided by the worse one, therefore, its value is always no less than 1).

From Table 4.13, we can observe a few things. Firstly, comparing number 1 and 2, 3 and 4,..., 23 and 24, the only difference between them is using o_rate in one case and $diss$ in the other. In all pairs, using o_rate always leads to bigger R^2 values. The last column "Pair comparison" presents the increase rate of R^2 when using o_rate

to replace *diss*. Therefore, we conclude that *o_rate* has more predictive ability than *diss*. Secondly, we may use *ave_p* and *dev* to replace *ratio* to predict if the fused result is better than the best of the two results, or replace *first* and *second* to predict the performance of the fused result. However, in both cases, the substitute is not as good as the original one though the difference is not big. Thirdly, the prediction is very poor when we use *first* and *second* to predict the performance of the fused result and use *ratio* to predict if the fused result is better than the best of the two results. Therefore, related results are not presented. On the other hand, *ave_p* and *dev* can be decently used in both situations.

Both *o_rate* and *diss* are used for the same purpose; it is interesting to investigate why *o_rate* has more predictive ability than *diss*. Because it seems that the calculation of *o_rate* is primitive and that of *diss* is more sophisticated. However, we notice that the ranking difference is not fully considered when calculating *diss*. Let us consider two results L_1 and L_2 and these two results have no common documents at all. For any pair of documents, there are three possibilities. The first is that both of them occur in L_1 but not L_2; we can decide their ranking in L_1 but not L_2. The second is that both of them occur in L_2 but not L_1; we can decide their ranking in L_1 but not L_2. The third is that the two documents are from different results, therefore, we cannot decide the ranking of them in both L_1 and L_2. In short, for any pair of documents, we cannot decide their ranking in both L_1 and L_2 at the same time. A score of 0.5 (the highest is 1) is assigned to *diss* between L_1 and L_2. Another case is L_1 and L_2 have the same documents but these documents are ranked very differently in L_1 and L_2, it is possible that these two results have a very high *diss* value (greater than 0.5). This is not reasonable because what we interest here is a certain number of top-ranked documents which are used for data fusion, and those documents that do not appear in component results are much less important than those that appear. Thus we hypothesize that is why *diss* is not as good as *o_rate* as an indicator of results similarity (dissimilarity). Besides, for three or more systems, it is still very straightforward to calculate *o_rate*, *ave_p*, and *dev*. However, how to calculate *diss* and *ratio* is not clear.

4.2.6 Some Further Observations

In Sections 4.2.1 through 4.2.6, we have discussed several aspects that affect data fusion. Overlap rate among component results is one of them. Here we ignore the other aspects and focus on the overlap rate and its effect on data fusion. We divide the possible range of overlap rate [0,1] into 20 intervals [0, 0.05], [0.05, 0.1],..., [0.95, 1], then we observe the percentage of the improvement on performance that the fused result can obtain compared with the average performance of the component results. Figures 4.1 and 4.2 show the curves of the percentage of improvement for Comb-Sum in TREC 6 and TREC 2001, respectively. In Figure 4.1 and 4.2, each curve is associated with a number, which is the number of results involved in the fusion. These two figures demonstrate that there is a strong relation between the overlap rate and the performance improvement percentage of data fusion. When the overlap rate increases, the performance improvement percentage decreases accordingly. Besides, the figures also demonstrate that the number of results has considerable effect on data fusion as well. The curves for CombMNZ, which are not presented, are very similar to that for CombSum.

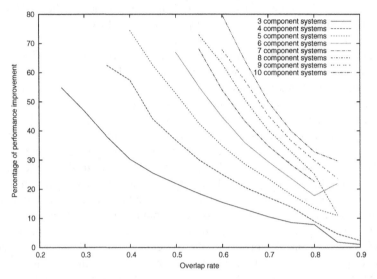

Fig. 4.1 Effect of overlap rate on the percentage of performance improvement (TREC 6, 3-10 results, CombSum)

Another observation is about the distribution of the overlap rate among component results. Figures 4.3 and 4.4 show the distribution of overlap rate among 3, 4,..., 10 results. In both figures, all curves should be well described by normal distribution curves. Also we observe the same pattern of curves in both figures when we have the same number of results. Another seemingly interesting phenomenon we observe is: the more systems we put in data fusion, the bigger values of overlap rate we obtain from the results of these systems. From 3 to 10 systems, the increase of overlap rate is considerable and monotonous. It suggests that there are a few systems which are quite different from each other (the average overlap rate for 3 systems is the lowest, around 0.5 in TREC 6 and around 0.6 in TREC 2001), but the number of quite different systems cannot be large. This phenomenon may be related to the factor that how overlap rate is defined. Refer to Equation 4.1. In this definition, no matter how many component systems there are, only those documents that occur in one of the results but not any others matter. If using another definition (e.g., refer to Equation 8.10), we might have different findings.

The relation between the number of component results and fusion performance is an interesting issue. As we know, the more component results are used, the greater the improvement we can expect for the fused results, if all other conditions are kept

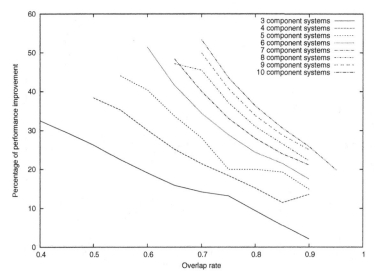

Fig. 4.2 Effect of overlap rate on the percentage of performance improvement (TREC 2001, 3-10 results, CombSum)

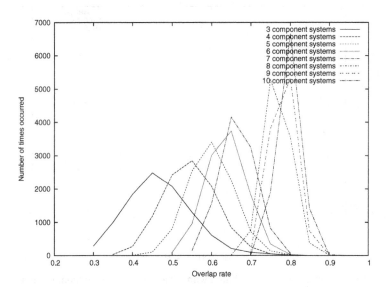

Fig. 4.3 Overlap rate distribution for 10,000 combinations in TREC 6

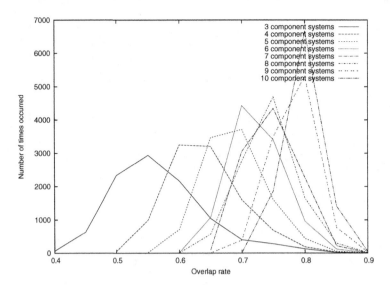

Fig. 4.4 Overlap rate distribution for 10,000 combinations in TREC 2001

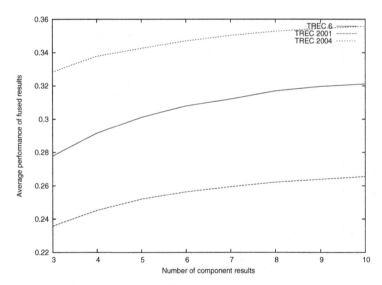

Fig. 4.5 Average performance of CombSum with a particular number of component results

the same. In Section 4.2.1 - 4.2.5, when using ln(*num*) to replace *num* along with other transformations, the prediction is more accurate in all cases. we also used both *num* and ln(*num*) in the regression analysis, but the effect (R^2) is just the same as using ln(*num*) only. This suggests that it is better to use a logarithmic function than a linear function to describe the relation between the number of component results and fusion performance. This is confirmed by Figure 4.5, in which the performance of CombSum is averaged for every certain number of component results. Each data value is the average of 10000 combinations over 50 (for TREC 6 and TREC 2001) or 249 (for TREC 2004) queries. For the three curves observed in TREC 6, TREC 2001, and TREC 2004, we use linear model and logarithmic model to estimate them by SPSS. For all three curves, the logarithmic functions for the estimation are at a significance level of .0000 (F = 326.9, 299.5, and 342.6 for TREC 6, TREC 2001, and TREC 2004, respectively); while the linear model for the estimation is not as good as the logarithmic model: significance = .0002 and F = 61.1 for TREC 6; significance = .0003 and F = 58.3 for TREC 2001; and significance = .0002 and F = 63.2 for TREC 2004.

4.2.7 Summary

In this section we have discussed the result of multiple regression analysis of three data fusion methods, CombSum, CombMNZ and Round-robin, with three groups of component results submitted to TREC 6, TREC 2001, and TREC 2004. Several different aspects, which are the number of component results, the overlap rate among the results, the average performance of the results, and the standard deviation of the average performance of the results are identified as highly significant variables that affect the performance of all three data fusion methods.

The analysis provides quite accurate prediction of the performance of the fused result with CombSum and CombMNZ. When using linear variables, all methods obtain a R^2 value between 0.778 and 0.852. The accuracy of the prediction can be improved by introducing nonlinear variables, and R^2 then ranges from 0.860 to 0.916. Especially using ln(*num*) to replace *num* (the number of results) can make considerable difference. When predicting the percentage of performance improvement of the fused result over the best component result, the prediction is quite accurate (R^2 values for all methods are between 0.811 and 0.863). Compared with Ng and Kantor's work [64] and Vogt and Cottrell's work [98, 99] (they focus on fusing two component systems and R^2 values are 0.204 and 0.06, respectively), the model discussed in this section is more useful for real applications.

Though a major goal of the analysis is to predict the performance of the fused result, the analytical result can also be used to predict if the fused result will be better than the best component result. Compared with Ng and Kantor's work [64], the analysis presented in this section is also more accurate. In all cases, a detection rate of 90% or over is observed; while in their work, about 70% of the detection rate is obtained for the testing runs and about 75% of the detection rate for the training runs.

Besides, their analysis only considered the case involving two component results, while the analysis discussed in this section is workable in a more general situation: more than 2 and variable numbers (3 to 10) of component results.

The experiment is conducted with 240,000 combinations in all, we can observe that almost all the fused results (using either CombSum or CombMNZ) are better than the average performance of component results, and some of the fused results (about 67%, using either CombSum or CombMNZ) are better than the best component result. Another interesting observation is the normal distribution of overlap rate among component results. This should be useful for us to improve the data fusion algorithms. The experiment also demonstrates that overlap-rate, one variable used in the model, has more predictive ability than the dissimilarity measure used in Ng and Kantor's work.

Finally, two variables, which are average performance of all component systems and standard deviation of the performances of all component systems, need document relevance judgements. For any given query, to obtain the exact values for these variables need to know the performances of all component systems, which demands document relevance judgements. It won't be realistic to do that each time for all component systems. Besides this, we may have two other options. The first is to evaluate the performances of all component systems using some training queries, and then we use these values from training queries for all test queries. The second is for every query, we estimate the performances of all component systems without document relevance judgements. Several methods (e.g., in ([2, 67, 88, 115]) on this issue have been proposed. With the help of these methods or alternatives, multiple regression analysis can be used to estimate the performances of component systems and also the fused result with less relevance judgment effort. However, further investigation is required on this.

4.3   Comparison of CombSum and CombMNZ

Many experiments have been conducted to compare data fusion algorithms using TREC data. Sometimes CombMNZ outperforms CombSum. Sometimes CombSum outperforms CombMNZ. People wonder which method is better and why. In this section, we are going to investigate this issue and make a comparison of them.

In the following we first apply statistical principles to data fusion. We show that CombSum is the best data fusion method statistically in the ideal situation: comparable scores and all component results are evenly distributed in the result space. Some further analysis is carried out to point out why sometimes CombMNZ outperforms CombSum and what can be done to improve the performance of CombSum. Empirical investigation is also carried out to confirm the conclusion.

4.3.1   Applying Statistical Principles to Data Fusion

In this subsection we first describe how the best way is to calculate scores according to statistical principles and some conditions needs to be satisfied accordingly [106].

Suppose we have a group of documents $\{d_1, d_2, ..., d_m\}$. For every d_j, we obtain n independent observation values, $s_{1j}, s_{2j}, ..., s_{nj}$, $(1 \leq j \leq n)$, for it. These observation values are estimated probabilities of this document being relevant to the information need. We define m statistical random variables $X_1, X_2, ..., X_m$. Each of them (X_j) is for a document d_j with n observation values. If all these observation values are independent, then the mathematical expectation of X_j can be calculated by the following equation

$$E(X_j) = \frac{1}{n} \sum_{i=1}^{n} s_{ij} \qquad (4.2)$$

All the documents can be ranked according to the mathematical expectations of these random variables. Actually, this is the case for fusing results from n information retrieval systems $\{ir_1, ir_2, ...ir_n\}$. Each of them contributes an observation value to each document d_j. Comparing Equation 4.2 with $\sum_{i=1}^{n} s_{ij}$ (see Equation 1.1), which is the equation used by CombSum, it is obvious that they are equivalent since a constant factor of $1/n$ in Equation 4.2 does not affect the rankings of all the documents involved.

In the long run, the above calculating method can bring us the most accurate estimation. It should be true that the method is better than the others if we consider enough cases. However, the best method statistically does not guarantee that in every single case the method is always better than the others.

More importantly, there are two conditions that need to be satisfied to guarantee the effectiveness of the method. Firstly, all observation values are probabilities of the documents being relevant to the query. Secondly, all observation values are randomly chosen for any variable X_j.

For the first condition, we can replace it with a weaker one: all observation values from different results are comparable with each other. Since we only concern the ranking of documents in the fused result, both of them are equal on this respective. As discussed at the beginning of Section 3, there are two types of comparability: comparability inside a result and comparability across results.

The second condition cannot be satisfied directly for the data fusion issue. Since each result is from a different information retrieval system, it is simply not possible for us to generate numerous information retrieval systems so as to obtain all different kinds of observation values needed. Instead of that, we can do something on the other side. Let us assume that we have a group of randomly selected results, then they must be evenly distributed in the whole result space. Now we can use this new condition " all the results are evenly distributed in the result space" to replace the original condition. This should be acceptable. A well-known example is to generate random numbers by computers. In order to generate real random numbers, a physical device must be used. However, in computer science, people are satisfied with using computers to generate pseudo-random numbers. The "random" numbers generated by computers are from an algorithm, therefore, they are not chosen randomly. But we can design an algorithm that makes the generated numbers to be evenly distributed in the desired range.

If the above two conditions are satisfied, then we can expect that CombSum performs better than others such as CombMNZ if a large scale experiment is carried out.

Finally, there are some further comments about the second condition. One is we may not be interested in the whole result space. If some results are very poor, then including them can only lead to poor fusion performance. On the other hand, we can only use a few available information retrieval systems to generate results. Due to the difficulties and costs incurred for developing new information retrieval systems (especially using new technologies), the number of independent sample results is limited and we may not have samples from certain part of the result space. Therefore, we may just focus on a sub-space that we have results in the neighbourhood.

The other one is the consideration for the operability of the second condition. As a matter of fact, to check if a group of results are evenly distributed in a sub-space needs some effort. We may use a stronger but simpler condition to replace it: every pair of the selected results have equal similarity. Let us see an example to illustrate this. Suppose that we have three results L_1, L_2, and L_3 and we can use a kind of distance (e.g., Euclidean distance) to measure the similarity of any pair of results. The condition of equal similarity means that we have $dist(L_1, L_2) = dist(L_2, L_3) = dist(L_1, L_3) = a$. These three results are surely evenly distributed in a certain space On the other hand, if L_1, L_2, and L_3 are on a straight line and $dist(L_1, L_2) = dist(L_2, L_3) = a$, but $dist(L_1, L_3) = 2a$, then these three results are evenly distributed in a sub-space, but the condition of "equal similarity between any pair of results" does not hold anyway.

4.3.2   Further Discussion about CombSum and CombMNZ

Both CombSum and CombMNZ were proposed by Fox and his colleagues [32, 33]. Actually, comparing the score calculating methods of CombSum and CombMNZ, which are $g(d) = \sum_{i=1}^{n} s_i(d)$ (Equation 1.1) and $g(d) = m * \sum_{i=1}^{n} s_i(d)$ (Equation 1.2) respectively. They bear some similarity. The difference between them is an extra factor m for CombMNZ, which is the number of results in which the document in question appears. For those documents retrieved by the same number of retrieval systems, their relative ranking positions in the merged results are always the same for both CombSum and CombMNZ. This determines that the merged results for CombSum and CombMNZ are very closely correlated with each other.

When documents are retrieved by different number of retrieval systems, CombSum and CombMNZ may rank them in different ways. CombMNZ favours those documents that are retrieved by more retrieval systems than CombSum does. For example, suppose we have two documents d_1 and d_2, d_1 is retrieved by 2 retrieval systems with a set of scores $\{0.6, 0.6\}$, while d_2 is retrieved by 3 retrieval systems with a set of scores $\{0.3, 0.2, 0.5\}$. Using CombSum, d_1 with a score of 0.12 will be ranked ahead of d_2 with a score of 0.10; while using CombMNZ, d_2 with a score of 3.0 will be ranked ahead of d_1 with a score of 2.4.

What can we learn from the above example? There are two things. First, if some of the component retrieval systems are more similar than the others, then they are very likely to assign similar scores to the same documents. CombMNZ is less affected than CombSum by this problem because it favours those documents that are retrieved by more component retrieval systems. In the above example, if two of three systems are very similar, then it is very likely that the same documents will be retrieved by both of them. For those documents that are retrieved by both of them but not the third one, their scores are over-estimated. ComMNZ can alleviate this problem by promoting the scores of those documents that are retrieved by all three retrieval systems. Of course, we do not mean that CombMNZ provides a prefect solution to this problem. There are many other alternative promotion policies. In Lee's experiment, he tried some other alternatives. Apart from using m, the number of results in which the document in question is retrieved, he used square root of m, square of m, etc. Although in Lee's experiment CombMNZ performed better than other alternatives, there is no guarantee that CombMNZ can always beat others in other occasions. The best promotion policy varies when we fuse different component results.

Second, score normalization does affect the performance of CombSum. The zero-one normalization method is commonly used for data fusion algorithms such as CombSum and CombMNZ. As discussed in Section 3.1.1, such a normalization method systematically over-estimates the top-ranked documents and under-estimates the bottom-ranked documents, since the top-ranked documents are not always relevant and the bottom-ranked documents are not always irrelevant (see Figure 3.1). Such a situation may be favourable to CombMNZ as well. Considering two documents d_1 and d_2, they obtain different numbers of non-zero scores and are ranked oppositely in CombSum and CombMNZ. Suppose d_1 obtains fewer non-zero scores than d_2, then it must be the case that the calculated score of d_1 is larger than that of d_2 in CombSum. Otherwise, both CombSum and CombMNZ will rank d_1 and d_2 in the same way: d_2 ahead of d_1. Comparing d_1 with d_2, d_1 has fewer non-zero scores but the sum of its scores is larger, and d_2 has more non-zero scores but the sum of its scores is smaller. It is very likely that some scores of d_1 are over-estimated (because they are larger), and some scores of d_2 are under-estimated (because they are smaller). Therefore, such an inaccuracy of normalized scores may cause CombSum to make a wrong decision, while CombMNZ is in a better position to avoid such errors. As a result, CombMNZ may outperform CombSum because of this. Having said that, there is no guarantee that CombMNZ can beat CombSum even in a quite favourable condition to CombMNZ, since the remedy provided by CombMNZ is far from perfect. This can be seen clearly later in Chapters 5 and 6.

In summary, comparable scores and evenly distributed results are two conditions for the best possible performance of CombSum. If either or both of them are not satisfied, then it is possible for CombMNZ to outperform CombSum.

4.3.3 Lee's Experiment Revisited

Let us revisit Lee's work in [48] because it is likely that is one of the best-known experiments. He used six retrieval results submitted to the TREC 3 ad-hoc track, namely *westp1*, *pircs1*, *vtc5s2*, *brkly6*, *eth001*, and *nyuir1* in his experiment. Since the raw scores provided by these component results were not comparable, the zero-one score normalisation method (Section 3.1.1) was used in the experiment for all the component results.

In his work, he defined pair-wise relevant overlap and pair-wise irrelevant overlap to measure dissimilarity between a pair of component systems. They are defined as:

$$RO(L_1, L_2) = \frac{2 * NR(L_1, L_2)}{NR(L_1) + NR(L_2)} \tag{4.3}$$

$$IO(L_1, L_2) = \frac{2 * NI(L_1, L_2)}{NI(L_1) + NI(L_2)} \tag{4.4}$$

Here $RO(L_1, L_2)$ denotes the relevant overlap rate between L_1 and L_2; $IO(L_1, L_2)$ denotes the irrelevant overlap rate between L_1 and L_2; $NR(L_1)$, $NR(L_2)$, $NR(L_1, L_2)$ denotes the number of common relevant documents in L_1, L_2, and both L_1 and L_2 respectively; while $NI(L_1)$, $NI(L_2)$, $NI(L_1, L_2)$ denotes the number of common non-relevant documents in L_1, L_2, and both L_1 and L_2, respectively.

For any pair from those six results, their relevant overlap rate is greater than their non-relevant overlap rate. He used it to explain why CombMNZ performed better than other methods including CombSum. The above hypothesis was disapproved by Beitzel et al. [10]. However, they did not explain what the real reason was.

Although the focus of Lee's experiment is to explain why data fusion methods such as CombMNZ and CombSum can bring effectiveness improvement, the experimental result shows that CombMNZ outperforms CombSum all the time, which is very impressive. It should be interesting to find out why this happens, since according to our analysis before, CombSum has the potential to outperform CombMNZ.

Actually, these six results are quite good in effectiveness and their performances measured by average precision are close, as shown in Table 4.14.

Table 4.14 Average precision of six results in Lee's experiment

brkly6	eth001	nyuir1	pircs1	vtc5s2	westp1
0.2775	0.2737	0.2722	0.3001	0.2914	0.3157

These results should also meet very well the requirement of evenly distributed samples in a result sub-space since these systems were developed by different participants. In order to confirm that, we calculate the overlap rate among these retrieved results as a way of evaluating similarity between a pair of results. See Equation 4.1 for the definition of overlap rate between two results. Different from Lee's methodologies of measuring $RO(L_1,L_2)$ and $IO(L_1,L_2)$, we do not distinguish relevant documents and irrelevant documents, but treat them equally. Table 4.15 shows the overlap rates of all pairs from a total number of six results.

Table 4.15 Overlap rates of all possible pairs from the six results in Lee's experiment

	eth001	nyuir1	pircs1	vtc5s2	westp1
brkly6	0.4677	0.3756	0.4150	0.3823	0.4052
eth001	1.0	0.4481	0.4365	0.3519	0.3813
nyuir1		1.0	0.4821	0.3804	0.4053
pircs1			1.0	0.3582	0.4153
vtc5s2				1.0	0.3588

In Table 4.15, the largest overlap rate is 0.4821 between *pircs1* and *nyuir1*, and the smallest one is 0.3519 between *eth001* and *vtc5s2*. all the overlap rates are quite close. We also calculate the Euclidean distance for all possible pairs of these six results. The zero-one method is used to normalize all the scores before the calculation process. In order to calculate the Euclidean distance, we need to define a union of the documents that are in one or both results:

$$u(L_1,L_2) = \{d_1,d_2,...,d_n\}$$

Suppose that for documents d_i, s_{1i} and s_{2i} are normalized scores that d_i obtain in L_1 and L_2, respectively. If d_i does not appear in a result at all, then a score of 0 is assigned. Then $L_1 = \{(d_1,s_{11}),(d_2,s_{12}),...,(d_n,s_{1n})\}$, $L_2 = \{(d_1,s_{21}),(d_2,s_{22}),...,(d_n,s_{2n})\}$, and we can use Equation 2.15 to calculate $dist(L_1,L_2)$. The Euclidean distances of all possible pairs from the six results are shown in Table 4.16.

Table 4.16 Euclidean distances of all possible pairs from the six results in Lee's experiment

	eth001	nyuir1	pircs1	vtc5s2	westp1
brkly6	23.81	25.28	24.68	25.26	24.93
eth001	0.00	24.55	24.89	26.44	25.98
nyuir1		0.00	23.99	26.02	25.56
pircs1			0.00	26.48	25.39
vtc5s2				0.00	25.76

In Table 4.16, the smallest distance is 23.81 between *brkly6* and *eth001*, and the largest is 26.48 between *pircs1* and *vtc5s2*. Although these two different methods of estimating similarity are not always consistent, for example, which pair are most similar, which pair are least similar, and so on, we can still conclude that similarity between all possible pairs are quite close no matter which method is used. Therefore, one of the two conditions is well satisfied, it is possible that the condition of comparable scores is not well satisfied and that makes CombSum not as good as CombMNZ.

Table 4.17 Performance (AP) for combining functions with different normalization settings

Methods	ways	(0,1)	(.02,1)	(.04,1)	(.06,1)	(.08,1)	(.1,1)
CombSum	2	0.3398	0.3407	0.3414	0.3419	0.3423	0.3426
CombMNZ	2	0.3430	0.3430	0.3428	0.3425	0.3423	0.3421
CombSum	3	0.3646	0.3658	0.3667	0.3674	0.3679	0.3683
CombMNZ	3	0.3685	0.3685	0.3682	0.3679	0.3675	0.3671
CombSum	4	0.3797	0.3808	0.3818	0.3824	0.3829	0.3831
CombMNZ	4	0.3835	0.3833	0.3828	0.3823	0.3817	0.3812
CombSum	5	0.3899	0.3909	0.3918	0.3924	0.3927	0.3927
CombMNZ	5	0.3927	0.3922	0.3916	0.3911	0.3905	0.3900
CombSum	6	0.3972	0.3985	0.3989	0.3990	0.3993	0.3992
CombMNZ	6	0.3991	0.3987	0.3982	0.3976	0.3971	0.3966

Therefore, we focus on the aspect of score normalization. As analysed in Section 3.1.1, there is room for the improvement of the zero-one linear normalisation method. Since not all top-ranked documents are relevant and not all bottom-ranked documents are irrelevant, 1 and 0 are not very good values for them. Instead of normalizing scores into the range of [0,1], as in Lee's experiment, we normalize scores by the fitting method. Scores are normalized into $[a, 1], (0 < a < 1)$. Note that we do not need to change the maximum limit 1 here. According to the principle of comparable scores, changing the minimum limit from 0 to a (varying) is enough for all possibilities, because what really matters is the ratio of the maximum limit to the minimum limit $(1/a)$, if ranking-based metrics are used. For example, normalizing scores into [0.1, 1] is equivalent to [0.05, 0.5] since 1/0.1 = 0.5/0.05.

Results are normalized with the fitting method with different a values. Table 4.17 shows the performance of fusion results using CombSum and CombMNZ with a few different a values (0.02, 0.04, 0.06, 0.08, 0.10). As in Lee's paper, a n-way data fusion presents a data fusion which has n component results. With 6 results, there are 15 2-way, 20 3-way, 15 4-way, 6 5-way, and 1 6-way combinations. Each data point in Table 4.17 is the average of all the combinations over 50 queries. A general trend is: the larger the a' value is, the better CombSum performs, but the worse CombMNZ performs. When $a = 0$, CombMNZ outperforms CombSum in all cases (which is exactly the same result observed in Lee's experiment); when $a = 0.1$, CombSum is better than CombMNZ with a [0.1,1] normalization and almost as good as CombMNZ with a [0,1] normalization.

Next we carry out another experiment. This time we normalize scores of 1000 documents into [0,1], but instead of taking all 1000 documents, we take fewer documents (100, 200, 400, 600, 800) for data fusion. In such a way we can get rid of those scores that are very close to zero. The fused results take the same number of documents as all component results do. This is another way of making the condition of comparable scores to be better satisfied. Table 4.18 shows the fused results[3]. The fewer documents we take for data fusion, the better CombSum performs compared with CombMNZ. When taking 1000 documents, CombMNZ always outperforms CombSum; When taking top 100 documents, CombSum is slightly better than CombMNZ in all cases.

The above two experiments show that score normalization does affect the effectiveness of CombSum. And the experiments also confirm our analysis for CombSum.

Table 4.18 Performance (AP) for combining functions with different number of documents (100, 200, 400, 600, 800, and 1000)

Methods	ways	100	200	400	600	800	1000
CombSum	2	0.2262	0.2801	0.3203	0.3353	0.3393	0.3398
CombMNZ	2	0.2261	0.2803	0.3208	0.3365	0.3413	0.3430
CombSum	3	0.2561	0.3123	0.3514	0.3618	0.3641	0.3646
CombMNZ	3	0.2258	0.3121	0.3514	0.3626	0.3662	0.3685
CombSum	4	0.2760	0.3331	0.3695	0.3769	0.3791	0.3797
CombMNZ	4	0.2756	0.3327	0.3691	0.3775	0.3811	0.3835
CombSum	5	0.2909	0.3480	0.3810	0.3873	0.3894	0.3899
CombMNZ	5	0.2903	0.3474	0.3804	0.3876	0.3908	0.3927
CombSum	6	0.3022	0.3594	0.3892	0.3952	0.3966	0.3972
CombMNZ	6	0.3016	0.3587	0.3886	0.3954	0.3971	0.3991

As discussed above, the six results in Lee's experiment are retrieved by six independently developed information retrieval systems. Therefore, They are well satisfied with one of the two conditions. In order to understand the effect of the other condition of even distribution, we add three more results, $eth002$, $nyuir2$, and $pircs2$, into the six-result group. Each of the three pairs, $eth001$ and $eth002$, $nyuir1$ and $nyuir2$, $pircs1$ and $pircs2$, were submitted by the same participant using the same information retrieval system but different settings or query representations. These three pairs' difference is much smaller than the other pairs. The overlap rate between $eth001$ and $eth002$ is 0.9357, the overlap rate between $nyuir1$ and $nyuir2$ is 0.9998, and the overlap rate between $pircs1$ and $pircs2$ is 0.7276. They are much larger than the average overlap rate of all other pairs: 0.4257.

Using these 9 retrieved results we form 9 combinations, and each of which includes 3 retrieved results and two of them are submitted by the same participant. Thus two of them is much similar to each other than the third one. We fuse these results using CombSum, CombMNZ, and two variants of CombMNZ. In CombMNZ (see Equation 1.2), the sum of scores from all component results is multiplied by a

---

[3] The experimental results presented here are different from Lee's experimental results. See Table 9 in Lee's paper [48]. One may obtain totally different conclusions from these two different experimental results. However, this does not necessarily mean that either of the experimental results must be wrong. The difference may be caused by the difference in score normalization, or the number of documents involved in the fused result. On the other hand, Lee only conducted experiments with 2, but not 3, 4, 5, and 6 component results.

factor m, which is the number of results in that the document is retrieved. We use $\sqrt{m}$ and $m * \sqrt{m}$ to replace m. Thus we obtain two variants, which are referred to as CombMNZ0.5 and CombMNZ1.5, respectively. The result is shown in Table 4.19.

Table 4.19 Average precision for fusing 3 results, in which two of them are much similar to each other than the third one

Group	CombSum	CombMNZ	CombMNZ0.5	CombMNZ1.5
eth001+eth002+brkly6	0.3139	0.3173	0.3167	0.3174
eth001+eth002+vtc5s2	0.3369	0.3434	0.3419	0.3438
eth001+eth002+westp1	0.3332	0.3378	0.3365	0.3379
nyuir1+nyuir2+brkly6	0.3285	0.3366	0.3343	0.3372
nyuir1+nyuir2+vtc5s2	0.3264	0.3255	0.3264	0.3242
nyuir1+nyuir2+westp1	0.3334	0.3418	0.3394	0.3431
pircs1+pircs2+brkly6	0.3338	0.3365	0.3361	0.3359
pircs1+pircs2+vtc5s2	0.3475	0.3509	0.3503	0.3508
pircs1+pircs2+westp1	0.3507	0.3536	0.3531	0.3533
average	0.3338	0.3382	0.3372	0.3382

In Table 4.19, we can see that for all combinations, CombMNZ and its two variants always outperforms CombSum with only one exception (CombSum and CombMNZ0.5 have equal performance when fusing $nyuir1$, $nyuir2$, and $vtc5s2$). It confirms that CombMNZ and its variants can alleviate the harmful effect of uneven distribution of component results.

One last question: how to improve the performance of CombSum in such a situation? One possible answer to this question is the linear combination method, which is a natural extension of CombSum. We will discuss how to assign weights in Chapter 5. The same issue will be addressed in Chapter 6 as well, but in a more formal style.

4.4 Performance Prediction of the Linear Combination Method

Predicting the performance of the linear combination method is more challenging than predicting the performances of CombSum and CombMNZ. Because the performance of the linear combination method is heavily affected by the weights assigned to the component results. Therefore, before we can predict anything, we need to assign certain weights to the results involved.

One possibility is: we might be able to predict the best possible performance if we can find the optimum weights for all the component results involved one way or another,

Vogt and Cottrell [98] did some research on this issue. They experimented with 61 runs submitted to TREC 5 (ad hoc track). Each fusion considered two results.

Therefore, there are a total of 1,830 result-level pairs or 36,600 query-level pairs (20 queries, #251-#270, were examined). Golden section search was used to find the optimised weights. For two results L_1 and L_2 (assuming that L_1 is more effective than L_2), Using four variables, which are p_1 (L_1's performance), p_2 (L_2's performance), $RO(L_1, L_2)$ (relevant overlap between L_1 and L_2, see Equation 4.3), and $IO(L_1, L_2)$ (irrelevant overlap between L_1 and L_2, see Equation 4.4), they predicted the performance of the linear combination method and the prediction was very accurate ($R^2 = 0.94$). More variables were also tried but no further improvement could be obtained. In those four variables, p_1 has the greatest impact on fusion performance with a normalized regression coefficient of 0.9366, which is followed by $RO(L_1, L_2)$ (0.1021), $IO(L_1, L_2)$ (-0.0581), and p_2 is the least one (-0.0228).

The research on this issue is far from complete. The situation of fusing three or more results needs to be considered. Other variables such as the number of component results and deviation of performance may also affect the performance of the linear combination method significantly. Prediction of performance improvement over the average performance or the best performance is also desirable.

Chapter 5
The Linear Combination Method

In data fusion, the linear combination method is a very flexible method since different weights can be assigned to different systems. When using the linear combination method, how to decide weights is a key issue. Profitable weights assignment is affected by a few factors mainly including performance of all component results and similarity among component results. In this chapter, we are going to discuss a few different methods for weights assignment. Extensive experimental results with TREC data are given to evaluate the effectiveness of these weights assignment methods and to reveal the properties of the linear combination data fusion methods.

5.1 Performance Related Weighting

The most profitable weighting schema for the linear combination method (LC) depends on a few factors of component results. But among those factors, performances of component results (especially the best one) have the strongest impact on appropriate weights assignment (see also Section 3.4). One weighting policy is to associate weights with performances.

For convenience, we rewrite Equation 1.3 as follows:

$$g(d) = \sum_{i=1}^{n} w_i * s_i(d) \tag{5.1}$$

$s_i(d)$ is the score obtained by d from ir_i and w_i is the weight assigned to system ir_i.

For each system ir_i, suppose its average performance (e.g., measured by average precision or recall-level precision) over a group of training queries is p_i, then p_i is set as ir_i's weight ($w_i = p_i$) in the simple performance-level weighting schema. The simple performance-level schema was used by quite a few researchers in their data fusion studies (e.g., in [3, 94, 113, 121]). In most cases (except in [94]) the linear combination method with the simple weighting schema outperformed both CombSum and CombMNZ.

S. Wu: Data Fusion in Information Retrieval, ALO 13, pp. 73–116.
springerlink.com   © Springer-Verlag Berlin Heidelberg 2012

Note that the performance of a retrieval system is always in the range of [0,1] when using AP or RP or other commonly used metrics. For the weighting schema ($w_i = p_i$), multiplying all the weights by a constant does not changing ranking, since all the weights are enlarged or reduced by the same times. On the other hand, using power functions of performance p_i, i.e., $w_i = p_i^{power}$ provides more options. When different powers are used, we can associate weight with performance in many different ways.

Let us take an example to illustrate this. Suppose we have two systems ir_1 and ir_2, measured by a certain metric, whose performance are $p_1 = 0.6$ and $p_2 = 0.8$, respectively. If a power of 1 is used, then the weights w_1 and w_2 of ir_1 and ir_2 are $p_1 = 0.6$ and $p_2 = 0.8$, respectively. The following table lists the normalized weights of them when a power of 0 to 5 is used. All the scores have been normalized using the equation $w_i = \frac{w_i}{w_1 + w_2}$ ($1 \le i \le 2$) to make the comparison more straightforward.

power	0	1	2	3	4	5
w_1	0.50	0.43	0.36	0.30	0.24	0.19
w_2	0.50	0.57	0.64	0.70	0.76	0.81
p_1/p_2	0.75	0.75	0.75	0.75	0.75	0.75
w_1/w_2	1.00	0.75	0.56	0.43	0.32	0.23

If a negative power is used, then the well-performed system is assigned a light weight, while the poorly-performed system is assigned a heavy weight. In such a case, we cannot obtain very effective results by fusing them using the linear combination method. Therefore, negative powers are not appropriate for our purpose and we only need to consider non-negative powers.

If a power of greater than 0 is used, then the well-performed system is assigned a heavy weight, while the poorly-performed system is assigned a light weight. Two special cases are 0 and 1. If 0 is used, then the same weight is assigned to all systems. Thus the linear combination method becomes CombSum. If 1 is used, then we obtain the simple performance-level weighting schema. If we use a very large power (much greater than 1) for the weighting schema, then the good system has a larger impact, and the poor system has a smaller impact on the fused result. If a large enough power is used, then the good system will be assigned a weight that is very close to 1, and the poor system will be assigned a weight that is very close to 0. Thus the fusion process will be dominated by the good component system and the fused result will be very much like the good component result. However, as we shall see later, using a large power like this is not the best weighting policy. In other words, it is possible that some appropriate weights can make the fused results better than the best component result.

In the above-mentioned example, we considered two component systems only. It is straightforward to expand the situation to include more than two component systems.

5.1.1 Empirical Investigation of Appropriate Weights

The purpose of the empirical investigation is to evaluate various weighting schemas and to try to find the most effective schemas. 4 groups of TREC data are used for the experiment [112]. Their information is summarized in Table 5.1 (See Appendix A for all the results involved and their performances measured by average precision). From Table 5.1 we can see that these 4 groups of results are different in many ways from track (web, robust, and terabyte), the number of results submitted (97, 78, 110, and 58) and selected (32, 62, 77, and 41) in this experiment [1], the number of queries used (50, 100, and 249), to the number of retrieved documents for each query in each result (1000 and 10000). They comprise a good combination for us to evaluate the data fusion methods. The zero-one linear normalization method was used for score normalization.

Table 5.1 Information summary of four groups of results (2001, 2003, 2004, and 2005) submitted to TREC and selected for the data fusion experiment

TREC group	2001	2003	2004	2005
Track	Web	Robust	Robust	Terabyte
Number of submitted results	97	78	110	58
Number of selected results	32	62	77	41
Number of queries	50	100	249	50
Number of retrieved documents	1000	1000	1000	10000
Average Performance (AP)	0.1929	0.2307	0.2855	0.3011
Standard deviation of AP	0.0354	0.0575	0.0420	0.0754

For all the systems involved, we evaluate their average performance measured by average precision over a group of queries. Then different values (0.5, 1.0, 1.5, 2.0,...) are used as the power in the power function to calculate weights for the linear combination method. In a year group, we choose m (m=3, 4, 5, 6, 7, 8, 9, or 10) component results from all available results for fusion. For each given m, we randomly select m component results from all available results for fusion. This process is repeated 200 times. Four metrics are used to evaluate the fused retrieval results. They are average precision (AP), recall-level precision(RP), percentage of the fused results whose performance in AP is better than the best component result (PFB(AP)), and percentage of the fused results whose performance in RP is better than the best component result (PFB(RP)). Besides the linear combination method with different weighting schemas, CombSum and CombMNZ are also involved in the experiment.

Tables 5.2-5.5 show the experimental result, in which four different powers (0.5, 1.0, 1.5 and 2.0) are used for the weighting calculation of the linear combination method. Each data point in the tables is the average of $8*200*|Q|$ measured values.

---

[1] Some results submitted include fewer documents than required (1000 or 10000). For convenience, those results are not used in the experiment.

Here 8 is the different number (3, 4,..., 9, 10) of component results used, 200 is the number of runs for a certain number of component results, and $|Q|$ is the number of queries in each year group (See Table 5.1).

Table 5.2 Performance (AP) of several data fusion methods (In LC(a), the number a denotes the power used for weight calculation; for every data fusion method, the improvement percentage over the best component result PIB(AP) is shown in parentheses)

Method	2001	2003	2004	2005
Best	0.2367	0.2816	0.3319	0.3823
CombSum	0.2614(+10.44)	0.2796(-0.71)	0.3465(+4.40)	0.3789(-0.89)
CombMNZ	0.2581(+9.04)	0.2748(-2.41)	0.3434(+3.46)	0.3640(-4.79)
LC(0.5)	0.2620(+10.69)	0.2841(+0.89)	0.3482(+4.91)	0.3857(+0.89)
LC(1.0)	0.2637(+11.41)	0.2865(+1.74)	0.3499(+5.42)	0.3897(+1.94)
LC(1.5)	0.2651(+12.00)	0.2879(+2.24)	0.3512(+5.82)	0.3928(+2.75)
LC(2.0)	0.2664(+12.55)	0.2890(+2.63)	0.3522(+6.12)	0.3952(+3.37)

Table 5.3 Percentage of the fused results whose performance in AP is better than the best component result (PFB(AP)) (In LC(a), the number a denotes the power used for weight calculation)

Method	2001	2003	2004	2005	Average
CombSum	83.18	54.62	87.56	50.81	69.05
CombMNZ	79.44	28.16	81.69	29.62	54.73
LC(0.5)	86.75	65.88	90.50	62.87	76.50
LC(1.0)	91.25	71.06	92.69	69.44	81.11
LC(1.5)	94.87	75.25	94.62	75.00	84.94
LC(2.0)	97.44	78.25	95.88	79.88	87.86

Table 5.4 Performance (RP) of several data fusion methods (In LC(a), the number a denotes the power used for weight calculation; for every data fusion method, the improvement percentage over the best component result PIB(RP) is shown)

Method	2001	2003	2004	2005
Best	0.2637	0.2977	0.3503	0.4062
CombSum	0.2815(+6.75)	0.2982(+0.17)	0.3629(+3.60)	0.4021(-1.01)
CombMNZ	0.2783(+5.54)	0.2943(-1.14)	0.3599(+2.74)	0.3879(-4.51)
LC(0.5)	0.2821(+6.98)	0.3009(+1.07)	0.3643(+4.00)	0.4077(+0.37)
LC(1.0)	0.2838(+7.62)	0.3024(+1.58)	0.3656(+4.37)	0.4112(+1.23)
LC(1.5)	0.2854(+8.23)	0.3034(+1.91)	0.3667(+4.68)	0.4137(+1.85)
LC(2.0)	0.2865(+8.65)	0.3043(+2.22)	0.3676(+4.94)	0.4156(+2.31)

Compared with the best component result, all the data fusion methods perform quite differently across different year groups. For all of them, TREC 2001 is the most successful group, TREC 2004 is the second most successful group, TRECs 2003 and 2005 are the least successful groups.

Table 5.5 Percentage of the fused results whose performance in RP is better than the best component result PFB(RP) (For LC(a), the number a denotes the power used for weight calculation)

Method	2001	2003	2004	2005	Average
CombSum	77.06	59.69	86.44	51.88	68.77
CombMNZ	73.37	49.50	80.81	30.56	58.56
LC(0.5)	80.50	68.94	89.31	63.38	76.06
LC(1.0)	86.44	73.06	91.75	69.94	80.30
LC(1.5)	89.81	76.13	93.69	75.25	83.72
LC(2.0)	92.44	79.25	95.37	79.88	86.74

Among all the data fusion methods, CombMNZ has the worst performance. CombMNZ does well in two year groups TRECs 2001 and 2004, but poorly in two other year groups TRECs 2003 and 2005. In fact, CombMNZ beats the best component result in TREC 2001 (+9.04% in AP, +5.54% in RP, 79.44 in PIB(AP), and 73.37 in PIB(RP)) and TREC 2004 (+3.46% in AP, +2.74% in RP, 81.69 in PIB(AP), and 80.81 in PIB(RP)), but worse than the best result in TREC 2003 (-2.41% in AP, -1.14% in RP, 28.16 in PIB(AP)), and 49.5 in PIB(RP) and TREC 2005 (-4.79% in AP, -4.51% in RP, 29.62 in PIB(AP), and 30.56 in PIB(RP)).

It is worthwhile to Compare CombMNZ with CombSum. CombSum is better than CombMNZ in all 4 year groups. However, just like CombMNZ, CombSum does not perform as well as the best component result in TRECs 2003 and 2005; while it performs better than the best component result in TRECs 2001 and 2004.

With any of the 4 powers chosen, the linear combination method performs better than the best component result, CombSum, and CombMNZ in all 4 year groups. However, the improvement rates of the linear combination method are different from one year group to another. Comparing all different weighting schemas used, we can find that the larger the power is used for weighting calculation, the better the linear combination method performs. The differences are especially larger for PIB(AP) and PIB(RP). Considering the average of all year groups, the percentage is over 76 for both PIB(AP) and PIB(RP) when a power of 0.5 is used. When the power reaches 2.0, the percentage is above 86 for both PIB(AP) and PIB(RP). This demonstrates that the linear combination method is more reliable than CombSum (69.05 for PIB(AP), 68.77 for PIB(RP)) and CombMNZ (54.73 for PIB(AP), 58.56 for PIB(RP)), when a power of 2 is used for weight calculation.

From the above experimental results we can see that the linear combination method increases effectiveness with the power used for weight calculation. Since only four different values (0.5, 1, 1.5, 2) have been tested, it is interesting to find how far this trend continues. Therefore, we use more values (3, 4, 5,... , 20) as power for the linear combination method with the same setting as before. The experimental result is shown in Figures 5.1-5.4.

In Figures 5.1 and 5.2, the curves of TREC 2004 reach their maximum as a power of 4 or 5 is used; while for the three other groups, the curves are quite flat and they reach their maximum when a power of between 7 and 10 is used. It seems that, for

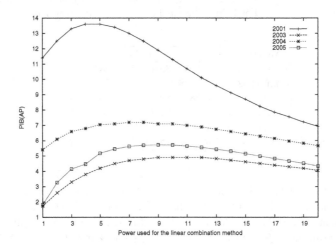

Fig. 5.1 Percentage of improvement (AP) of the linear combination method over the best component result when using different powers

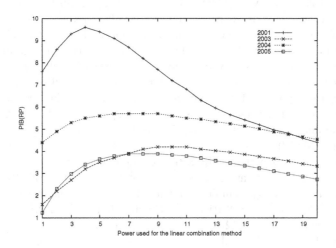

Fig. 5.2 Percentage of improvement (RP) of the linear combination method over the best component result when using different powers

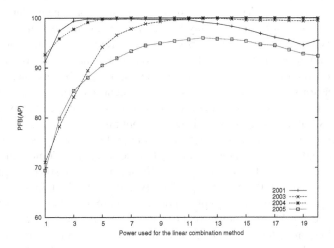

Fig. 5.3 Percentage of the fused results whose performance (AP) is better than the best component result for the linear combination method using different powers

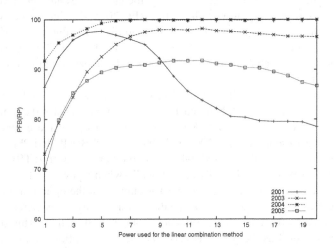

Fig. 5.4 Percentage of the fused results whose performance (RP) is better than the best component result for the linear combination method using different powers

obtaining the optimum fusion results, different powers may be needed for different sets of component results. This may seem a little strange, but one explanation for this is: data fusion is affected by many factors such as the number of component results involved, performances and performance differences of component results, dissimilarity among component results, and so on [123]. Therefore, it is likely that the optimum weight is decided by all these factors, not just by any single factor, though the performance of component results is probably the most important one among all the factors. Anyway, if we only consider performance, then a power of 1, as the simple weighting schema does, is far from optimum.

Two-tailed t tests is carried out to compare the performance of the data fusion methods involved [2]. The test shows that the differences between CombSum and the linear combination method (with any power 1-20) are always statistically significant at a level of .000 ($p < 0.0005$, or the probability is over 99.95%). In TREC 2001, the linear combination method is not as good as CombSum when a power of 12 or more is used for AP (or a power of 11 or more for RP). In all other cases, the linear combination method is better than CombSum. We also compare the linear combination method pairs with adjacent integer powers such as 1 and 2, 2 and 3, etc. In every year group, most of the pairs are different at a level of .000 with a few exceptions. In TREC 2001, when comparing the pair with powers 4 and 5, the significance value is .745 (AP). In TREC 2003, the significance value is .024 (AP) for the pair with powers 10 and 11. In TREC 2004, the significance value is .037 (RP) for the pair with powers 6 and 7; the significance values are .690 (AP) and .037 (RP) for the pair with powers 7 and 8. In TREC 2005, the significance value is .072 (PP) for the pair with powers 7 and 8; the significance value is .035 (RP) for the pair with powers 8 and 9; the significance value is .002 (AP) for the pair with powers 9 and 10. All such exceptions happen when the linear combination method is at the peak of performance.

In Figures 5.3 and 5.4, PFB(AP) and PFB(RP) increase very rapidly with the power at the beginning. Both of them reach their maximum almost at the same point as corresponding AP and RP curves. After that, all the curves decrease gently with power. In two year groups TRECs 2001 and 2004, both PFB(AP) and PFB(RP) are around 90 when a power of 1 is used; while in two other year groups TRECs 2003 and 2005, both PFB(AP) and PFB(RP) are about 70 when a power of 1 is used.

If we consider all the metrics and all year groups, then the optimum points are not the same. However, we can observe that the performance always increases when the power increases from 1 to 4 for all year groups and all metrics. This suggests that using 2 or 3 or 4 as the power is very likely a better option than using 1, as the simple weighting schema does. Compared with the simple weighting schema, an improvement rate of 1% to 2% is achieved either measured by AP or RP, and an improvement rate of 10% to 15% is achieved on PFB(AP) and PFB(RP) by using a power of 4. TREC 2001 reaches its peak when a power of 4 is used. However, the three other groups continue to increase for some time. For example, if we use a power of 7 or 8, and compare it with the simple weighting schema, then an

---

[2] The linear combination method is regarded to as being multiple methods when different powers are used.

improvement rate of 2.5% on AP and RP, and an improvement rate of 15% to 20% on PFB(AP) and PFB(RP) are achievable.

5.1.2   Other Features

In this section we discuss some related features of the linear combination data fusion method by empirical investigation. The same data sets as in Section 5.1.1 are used. First let us see the effect of the number of component results on the fused results. For each group of component results chosen randomly, we fuse them using the linear combination method 20 times, each with a different weight (a power of 1, 2, ,... , 20 is used). Then we choose the best one from all 20 fusion results generated for consideration. The experimental result is presented in Figure 5.5. As in Section 5.1.1, each data point is the average of $8*200*|Q|$ measured values, where 8 is the number of different groups (including 3, 4,...., 10 component results), 200 is the number of runs with a specific number of component results, and $|Q|$ is the number of queries in each year group. In this subsection, we only present results measured by AP. Similar results are observed using RP.

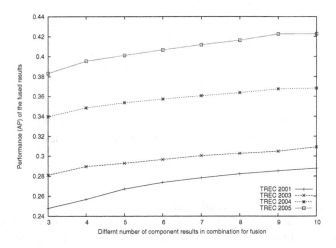

Fig. 5.5 The relationship between the number of component results and the performance of the fused result measured by AP

In Figure 5.5 we can see that all 4 curves almost increase linearly with the number of component results. It demonstrates that the number of component results has a positive effect on the performance of the fused result using the linear combination method. Comparing Figure 5.5 with Figure 4.5, we can see that the number of component systems has larger positive impact on the fusion performance of the linear combination method (linearly) than on that of CombSum (logarithmically).

Now let us take a look at the relationship between the average performance of component results and the improvement rate of performance that the fused result can obtain over the average performance. For a group of component results, we chose the best fusion result as above. Then we divide the average performance measured by AP into equal intervals 0-0.0999, 0.1000-0.1099, ..., put all the runs into appropriate intervals, and calculate the improvement rate of performance of the fused results for each interval.

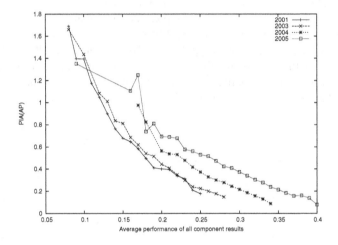

Fig. 5.6 The relationship between the average performance (measured by AP) of component results and the improvement rate of the fused results over average performance of component results

Figure 5.6 shows the experimental result. Each data point is the average of all the runs in an interval and "0.1" on the horizontal axis denotes the interval of 0.1000-0.1099 and so on. In all year groups, the rate of performance improvement of the fused results decrease rapidly with the average performance of component results, with a few exceptions. All the exceptions occur in those data points that include very few runs. This demonstrates that the better the component results are in performance, the less benefit we can obtain from such a data fusion method.

When using the linear combination method to fuse results, the contribution of a component result to the fused result is determined by the weight assigned to it. It is interesting to observe the contribution of each component result to the fused result when different weighting schemas are used. In order to achieve this, we carry out an experiment as follows: first, in a year group, a given number of component results are chosen and fused using different power weighting schemas; second, the best and the worst component results are identified; third, for each weighting schema, we calculate the Euclidean distance (see Equation 2.15) between the worst and the best component results, between the worst component results and the fused results, and between the best component results and the fused results after normalizing them using the zero-one normalization method.

Using the same data as in Section 5.1.1, we carry out the experiment and hereby present the result of TREC 2001 in Figure 5.7. The results for the three other year groups are analogous and the figures are omitted.

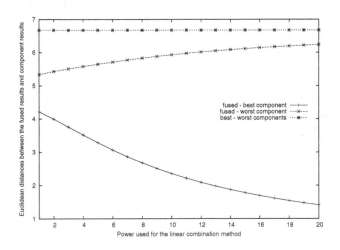

Fig. 5.7 The Euclidean distances between the fused and the best component results, the fused and the worst component results, and the best and the worst results in TREC 2001

In Figure 5.7, each data point is the average of 8 different numbers (3, 4., ..., 8) * 200 combinations * 50 queries. Since the Euclidean distance between the worst and best component results is not affected by data fusion at all, it is a flat line in Figure 5.7. As expected, when the power increases, the distance between the worst component results and the fused results increases, while the distance between the best component results and the fused results decreases. However, quite surprisingly, even when the power is as high as 20, the distance between the best component results and the fused results is about 1.5 (compared with about 6, the distance between the worst component results and the fused results), which is not close to 0 by any means. This demonstrates that even when a power as high as 20 is used, the fused results are not just affected by the best component results. Other component results (apart from the best ones) still have significant impact on the fused results.

One phenomenon in TREC may be worth investigating in this study. In each
year group, any participant may submit more than one run to the same track. Those
runs are more similar than usual since many of them are just obtained by using the
same information retrieval system but different parameter settings/different query
formats. Up to this point, we do not distinguish runs from the same participant or
not. In each combination, hence a few component results may come from the same
participants, the fused results may be biased to them. In order to avoid such things
from happening, we divide participating runs into sets and all the runs submitted
by the same participant are put in the same set. We randomly select runs from all
the sets as before. But for any single combination, we choose at most one run from
any particular set. We refer this to be the non-same-participant restriction. We use
TRECs 2001 and 2003 in this experiment. As before, in a year group, we choose m
(m=3, 4, 5, 6, 7, 8, 9, or 10) component results from all available results for fusion
under the non-same-participant restriction. For each setting of m, we form and test
200 combinations. Figure 5.8 shows the experimental result.

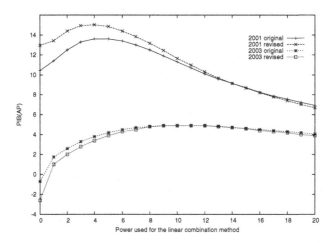

Fig. 5.8 Comparison of fusion performance of two groups, the non-same-participant restriction is applied to one group ("2001 revised" and "2003 revised") and no such restriction
to the other group ("2001 original" and "2003 original") for comparison, the measure used
is the percentage of improvement (in AP) of the linear combination method over the best
component result

In Figure 5.8, "2001 revised" and "2003 revised" denote the groups that observe
the non-same-participant restriction, while "2001 original" and "2003 original" denote the groups which do not observe the non-same-participant restriction[3]. The

---

[3] They are the curves labelled as "2001" and "2003" in Figure 5.1.

curves of "2001 original" and "2001 revised" are not always very close, this is because different component results are involved in the two corresponding groups, though it is the same case for "2003 original" and "2003 revised". We can see that the two corresponding curves for comparison are the same in shape and they reach the maximum points with more or less the same power values. Therefore, it demonstrates that the even or uneven dissimilarity among component results does not affect our conclusion before.

In Section 5.1.1, the same group of queries are used for obtaining performance weights and evaluation of data fusion methods. Now let us investigate a more practical scenario: different groups of queries are used for obtaining performance weights and evaluation of data fusion methods. In order to do this, we divide all the queries in every year group into two sub-groups: odd-numbered queries and even-numbered queries. We obtain two different performance weights from these two sub-groups. Then for those odd-numbered queries, we evaluate all data fusion methods using the weights obtained from even-numbered queries; for those even-numbered queries, we evaluate all data fusion methods using the weights obtained from odd-numbered queries. This is referred to as two-way across validation [61]. Three year groups (TREC 2001, 2003, and 2004) are used in this experiment. All the combinations involved are the same as in Section 5.1.1. Thus we can carry out a fair comparison of these two different settings.

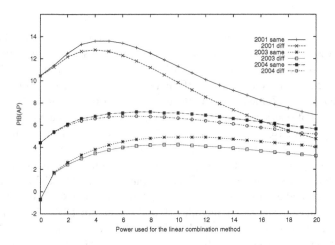

Fig. 5.9 Comparison of fusion performance in two different settings, one setting uses the same group of queries for obtaining performance weights and evaluation of data fusion methods and the other setting uses different group of queries for obtaining performance weights and evaluation of data fusion methods, the measure used is the percentage of improvement (on AP) of the linear combination method over the best component result

Figure 5.9 shows the result. "2001 diff", "2003 diff", and "2004 diff" are from the setting that uses different group of queries for the training of performance weights and evaluation of fusion methods; while "2001 same", "2003 same", and "2004 same" are from the setting that uses the same group of queries for the training of performance weights and evaluation of fusion methods[4].

From Figure 5.9 we can see that using the same group of queries can always lead to better fusion performance than using different groups of queries for the training of performance weights and evaluation of fusion methods. The difference between the two corresponding curves becomes larger when a larger power is used for weight assignment. However, the two corresponding curves in each year group bear the same shape. Therefore, our observations and conclusions obtained before by using the same group of queries for performance weights training and data fusion methods evaluation also hold for the situation that different queries are used for the training of performance weights and evaluation of fusion methods.

Table 5.6 Correlations of performances between two sub-groups of queries in each year group

Groups	Pearson	Kendall's tau_b	Spearman
TREC 2001	0.856	0.532	0.685
TREC 2003	0.961	0.720	0.880
TREC 2004	0.957	0.798	0.938

In Figure 5.9, we can also find that, in each year group, the difference between the two curves are different. Among them, TREC 2001 has the biggest difference, TREC 2004 has the smallest difference, while TREC 2003 is in the middle. This is due to the different accuracies of performance estimation for these three groups of component results. Table 5.6 shows the correlation of performances between the two sub-groups of queries in each year group. Among these three groups, the correlation between the two sub-groups in TREC 2004 is the strongest, which is followed by the two sub-groups in TREC 2003, and the correlation between the two sub-groups in TREC 2001 is the weakest. Further, we guess this is mainly due to the different numbers of queries used in each year group. In TREC 2001, there are only 50 queries, and each sub-group includes 25 queries; in TREC 2003, there are 100 queries, and each sub-group includes 50 queries; and in TREC 2004, there are 249 queries, and each sub-group includes 125 or 124 queries. This experiment also suggests that a relatively large number of queries (at least 50) should be used for an accurate estimation of performance.

---

[4] They are the curves labelled as "2001", "2003", and "2004" in Figure 5.1.

5.2 Consideration of Uneven Similarity among Results

In Section 5.1, the discussion is based on the assumption that the performance of component results is the sole factor that affects the performance of the fused result. As a matter of fact, apart from performance, similarity among component results is another significant factor. If all component results are very similar or strongly corre-lated (it is very likely the case for several retrieval models, several query formations, and so on in the same information retrieval system, as suggested by Beitzel et. al.'s experimental results [10]), then no effective remedy is available since all the results are very much alike. The best thing we can do is to avoid this situation from occur-ring. If some of the component results are more strongly correlated than the others, then the opinion among the strongly correlated results will dominate the data fusion process. Such a phenomenon is potentially harmful to data fusion, since highly cor-relating results are very likely to over-promote some common irrelevant documents among them. In this section, we are going to discuss some effective data fusion methods which can solve this problem [121]. The general idea is to assign corre-lation weights to all results or/and their combinations; then the linear combination method is used in the fusion process. These methods are referred to as correlation methods.

5.2.1 Correlation Methods 1 and 2

First let us see how to calculate the strength of the similarity/correlation between two component results. Quite a few different methods such as overlap rate and Spear-man's rank coefficient can be used for this purpose. See Equation 4.1 for the defini-tion of overlap among a group of component results. More specifically, if there are two results L_i and L_j, then the overlap rate between L_i and L_j can be defined as

$$o\_rate_{ij} = \frac{2 * |L_i \wedge L_j|}{|L_i| + |L_j|} \tag{5.2}$$

Here $|L_i|$ denotes the number of documents in L_i. Suppose we have n results L_1, L_2,..., L_n, then for every result L_i, we obtain its average overlap coefficient with all other results

$$w_i = 1 - \frac{1}{n-1} \sum_{j=1,2,...,n, j \neq i} o\_rate_{ij} \tag{5.3}$$

In such a way, if a component result correlates more strongly with all other results on average, then a lighter weight is assigned to it; otherwise, a heavier weight is assigned to it. Having w_i for every result L_i, we use the linear combination method (see Equation 1.3) to fuse all the results: $g(d) = \sum_{i=1}^{n} w_i * s(d)$

The above-mentioned method is referred to as correlation method 1. If a result is very independent, then it is very likely to obtain a lower overlap rate than some other results, and this again leads to a heavier weight assigned to it and make the result more influential to data fusion. Conversely, if a result is very dependent, then it is very likely to obtain a higher overlap rate, and this again leads to a lighter weight assigned to it and makes it less influential to data fusion. Using this method, we do not need other information except for the component results as in CombSum and CombMNZ.

In the above-mentioned method, the weight of every result is calculated in every query, which may vary from one query to another randomly. A moderated method to this (we call it correlation method 2) is to calculate the weights of all results by using a group of training or historical query results and take the average values of them.

Suppose there are n results, and each result includes m documents, then calculating the overlap rate between any pair of results needs $O(m)$ time if we use a hash table. There are $n*(n-1)/2$ possible pairs of results for a total of n results, and to calculate the overlap rates among all possible pairs takes $O(m*n^2)$ time, which is the time complexity of correlation methods 1 and 2 for weights assignment. With weights assigned, the complexity of the data fusion process is the same as that of the linear combination method ($O(m*n)$).

5.2.2   Correlation Methods 3 and 4

The general idea of correlation methods 3 and 4 is the same as that of correlation methods 1 and 2. The only difference is that we use a different method for weights assignment.

The weights assignment process is: first we merge all n results $L_1, L_2,..., L_n$, into a single set of documents S. For every document d in S, a count number is used to indicate the number of results which includes d. Furthermore, we also say that every corresponding document of d in $L_1, L_2, ..., L_n$ has the same count number as d has.

Next, we sum up the count numbers of all documents for every result. Suppose that all results include m documents for the same query. For result L_i, if the sum of count number of all documents in L_i is m, which means that all documents in this result are totally different from the others and no overlap exists, we assign a weight of 1 to L_i; if the sum is $m*n$, which means that all n results just include the same documents, we assign a weight of $1/n$ to L_i; if the sum v is between m and $m*n$, we assign a weight of $1 - \frac{v-m}{m*n}$ to L_i.

The above-mentioned weighting calculation can be carried out in every query and for that query only (correlation method 3). An alternative is keeping all historical weights and using their average for the forthcoming queries (correlation method 4).

Because the overlap rates of all n results (each with m documents) can be calculated by one scan of each, the time complexity of the weights assignment process is $O(m*n)$. The data fusion process of these two methods takes $O(m*n)$ time as the linear combination method does.

5.2.3 Correlation Methods 5 and 6

For n results L_i $(1 \leq i \leq n)$, correlation may exist among multiple results. In these two methods, we assign a weight to every possible combination of the n results. For n results, there are $2^n - 1$ combinations in total. For example, if we have 3 results L_1, L_2, and L_3, then there are $2^3 - 1 = 7$ combinations, which are $\{L_1\}$, $\{L_2\}$, $\{L_3\}$, $\{L_1,L_2\}$, $\{L_1,L_3\}$, $\{L_2,L_3\}$, and $\{L_1,L_2,L_3\}$. We can use c_j $(1 \leq j \leq 2^n - 1)$ to represent any of these combinations. There is a convenient way to decide the correspondence between a number j and the combination that c_j represents: we represent j in binary format with n digits. If there is a digit 1 in place i (counting from right to left), then L_i will be included in the combination; if 0 appears in place i (counting from right to left), then L_i will not be included in the combination. For three results, every c_j and its corresponding combination is listed as below:

- c_1 $\{L_1\}$ (binary format of 1 is 001)
- c_2 $\{L_2\}$ (binary format of 2 is 010)
- c_3 $\{L_1,L_2\}$ (binary format of 3 is 011)
- c_4 $\{L_3\}$ (binary format of 4 is 100)
- c_5 $\{L_1,L_3\}$ (binary format of 5 is 101)
- c_6 $\{L_2,L_3\}$ (binary format of 6 is 110)
- c_7 $\{L_1,L_2,L_3\}$ (binary format of 7 is 111)

This combination representation is useful for merged documents as well. For every document d in the merged set, very often we need to indicate all the results in which the document is included. It is a combination of the results, and the subscript number j $(1 \leq j \leq 2^n - 1)$ in c_j is referred to as the inclusion number of the merged document. In the following, we define two functions. The first one is $binary(i,j)$, which checks if a 1 or 0 appears in place i of j's binary format (from right to left). If 1 appears in place i of j's binary format, then $binary(i,j)=1$; if 0 appears in place i of j's binary format, then $binary(i,j)=0$. This function is used to decide if a particular result is included in a particular combination. The second one is $count(j)$, which counts the number of 1's in j's binary format. This function is used to calculate the number of results included in a particular combination. For example, $binary(1,5) = 1$; $binary(1,4) = 0$; $count(11) = 3$.

We assign a weight w_j to every combination c_j by a group of results from training queries or the current query under processing. For these weights, the condition

$$\sum_{j=1}^{2^n-1} w_j * binary(i,j) = 1$$

must be satisfied by all the results L_i $(1 \leq i \leq n)$. This guarantees that all the weights are normalized. If we have 3 results, then $w_1 + w_3 + w_5 + w_7 = 1$ (for result 1), $w_2 + w_3 + w_6 + w_7 = 1$ (for result 2), and $w_4 + w_5 + w_6 + w_7 = 1$ (for result 3) must be satisfied.

With all these weights, we are able to fuse results. For any document d, we assume that it obtains n scores $(s_1(d), s_2(d),...,s_n(d))$ from n results. If d is not included in L_i, then we simply assign a score of 0 to $s_i(d)$. Then we use the following formula to calculate the global score of d:

$$g(d) = \sum_{j=1}^{j=2^n-1} w_j * \sum_{i=1}^{n} \frac{(s_i(d) * binary(i, j))}{count(j)}$$

Next, let us discuss how to assign these weights by training or on the fly. Suppose we obtain n results, all of which include the same number (m) of documents, from all component systems for a given query. We can use the following algorithm to assign weights to w_j $(1 \leq j \leq 2^n - 1)$.

Algorithm 5.1

01   Input: n results L_i $(1 \leq i \leq n)$ from n different systems and each L_i includes m documents.
02   Output: w_j $(1 \leq j \leq 2^n - 1)$.
03   Combine n results into a single set T;
04   For every $d \in T$, indicating the results in which d is included using a inclusion number j;
05   Initialise w: assign 0 to every w_j;
06   For every element in the combined set, if its inclusion number is j, then let $w_j :=$ $w_j + 1$;
07   Normalizing w_j: for every w_j, let $w_j := w_j/m$.

Fig. 5.10 Weights assignment for correlation methods 5 and 6

Again, we have two options here for the above-mentioned weights assignment process: to carry it out in every query and use it for that query only (correlation method 5), or to keep all historical data and use their average for the forthcoming queries (correlation method 6).

The time complexity of correlation method 5 and 6 are $O(m * n + 2^n)$ for weights assignment, because it needs one scan of all results and all $2^n - 1$ weights needs to be assigned. For these two methods, the time complexity of the data fusion process is variable. One extreme situation is that all n results include the same collection of documents. Then we have m merged documents, and each of these documents has to use all $2^n - 1$ weights. Therefore, in such a situation the time complexity is $O(m * 2^n)$. If there is no or very little overlapping documents, then it only takes $O(m * n)$ time as the linear combination method.

In summary, all six correlation methods presented above are generalizations of CombSum. Correlation methods 1-4 become CombSum when all results obtain equal weights. Correlation methods 5 and 6 are more precise than the 4 other correlation methods because all types of possible correlation (e.g., correlation among more than two results) are considered.

5.2.4 Empirical Study

Six groups of results, which were the top 10 results[5] submitted to TREC 6 (ad-hoc track), 7 (ad-hoc track), 8 (ad-hoc track), 9 (web track), 2001 (web track), and 2002 (web track). were used. RR (Reciprocal Rank) is used in the evaluation of the experimental results.

Some statistics of these six year groups are presented in Table 5.7 (See also Appendix B for the titles and performances of all the runs involved). Among all these 6 year groups, the TREC 8 group is the best with an average of 0.8416 in performance, while the TREC 2002 group is the worst, whose average performance is only 0.4571.

Table 5.7 Some statistics of six groups of results (TRECs 6-9, TRECs 2001-2002)

Group	Average RR	Deviation of RR	Average overlap rate	Deviation of overlap rates
TREC 6	0.7672	0.0719	0.3777	0.1309
TREC 7	0.8196	0.0247	0.4414	0.1142
TREC 8	0.8416	0.0416	0.4925	0.1274
TREC 9	0.6247	0.0504	0.5185	0.1542
TREC 2001	0.6637	0.0422	0.5209	0.1396
TREC 2002	0.4571	0.0138	0.5332	0.2023

Table 5.8 Summary of performance comparison of data fusion methods (3-8 results, 957 combinations in TREC 6, 7, 8, 9, 2001 and 2002, 50 queries for each combination)

Group	RR	PIB(RR)	PFB(RR)
Best	0.7457	-	-
CombMNZ	0.7682	3.20	73.45
CombSum	0.7689	3.34	71.82
Cor1	0.7707	3.53	74.48
Cor2	0.7708	3.51	76.09
Cor3	0.7704	3.54	73.86
Cor4	0.7698	3.42	73.06
Cor5	0.7712	3.57	75.64
Cor6	0.7712	3.60	75.69

Out of a total of 10 results, there are 45 pairs of different combinations in each group. The overlap rates of all these pairs were measured and averaged. These averages are presented in the column "Average overlap rate" in Table 5.7. Considerable overlap exists in all cases. For all 6 groups, average overlap rates vary from 0.3773 to 0.5332, increasing monotonously year by year.

---

[5] For convenience, all selected results include 1000 documents for each query.

Table 5.9 Performance comparison of a group of data fusion methods (3-8 systems, 50 queries for each combination)

Method	Measure	3 Systems 120*6	4 Systems 210*6	5 Systems 252*6	6 Systems 210*6	7 Systems 120*6	8 Systems 45*6
CombMNZ	RR	0.7537	0.7628	0.7686	0.7728	0.7777	0.7824
	PIB(RR)	2.98	3.16	3.19	3.17	3.39	3.73
	PFB(RR)	71.58	74.64	74.24	73.78	71.35	72.46
CombSum	RR	0.7541	0.7629	0.7691	0.7737	0.7768	0.7816
	PIB(RR)	3.04	3.24	3.37	3.42	3.48	3.66
	PFB(RR)	70.34	73.37	72.59	71.71	69.40	71.01
Cor1	RR	0.7548	0.7650	0.7715	0.7759	0.7813	0.7830
	PIB(RR)	3.10	3.43	3.56	3.62	3.93	3.68
	PFB(RR)	71.21	75.75	76.22	75.43	72.31	68.84
Cor2	RR	0.7559	0.7655	0.7713	0.7760	0.7803	0.7836
	PIB(RR)	3.20	3.48	3.51	3.56	3.71	3.78
	PFB(RR)	73.61	76.78	77.81	76.00	74.12	75.73
Cor3	RR	0.7546	0.7649	0.7707	0.7757	0.7808	0.7846
	PIB(RR)	3.08	3.45	3.50	3.60	3.89	4.08
	PFB(RR)	70.94	76.07	75.20	74.25	71.35	69.20
Cor4	RR	0.7550	0.7647	0.7707	0.7756	0.7799	0.7735
	PIB(RR)	3.14	3.44	3.50	3.60	3.74	1.76
	PFB(RR)	71.49	76.64	75.43	74.17	71.90	54.37
Cor5	RR	0.7552	0.7654	0.7721	0.7767	0.7811	0.7841
	PIB(RR)	3.16	3.48	3.63	3.68	3.83	3.89
	PFB(RR)	71.90	75.75	77.87	76.86	74.52	69.93
Cor6	RR	0.7555	0.7654	0.7721	0.7766	0.7811	0.7840
	PIB(RR)	3.21	3.49	3.66	3.68	3.82	3.80
	PFB(RR)	72.84	75.84	77.35	75.61	75.93	72.91

Different number $\{3, 4, 5, 6, 7, 8\}$ of results for each data fusion run were used. Out of 10 results, there are 120, 210, 252, 210, 120, and 45 different combinations for 3, 4, 5, 6, 7, and 8 results, respectively. Data fusion runs were performed with all possible combinations in all 6 year groups. Eight data fusion methods were used: CombSum, CombMNZ, and correlation methods 1-6 (Cor1, Cor2, ..., Cor6). The average performances of these data fusion methods are presented in Table 5.8 for all 6 year groups (all possible combinations of 3-8 results out of 10 results, 5742 in total). The figures in the "RR" row indicate the performance (in RR) of the fused result using a particular data fusion method, while the figures in the "PIB(RR)" row indicate the percentage of improvement in performance of the data fusion method over the best result involved, and the figures in the "PFB(RR)" row indicate the percentage of runs at which the fused result is better than the best result involved.

For all the methods, their average performances are very close, and so are the percentages of improvement over the best result, which is between 3.20 and 3.60 for all the methods involved. A two-tailed, paired-samples t test was done to compare the means of the best result and of the fused results. It shows that for all 8 fusion methods their performances are very significantly better than that of the best component result at a significance level of .000. The performances of all data fusion methods involved in the experiment were also compared. It shows that CombMNZ is worse than CombSum at a significance level of .016, while both CombSum and CombMNZ are worse than all six correlation methods at a significance level of .000. In six correlation methods, method 5 and 6 are better than the four other methods at a significance level of no more than .016 (many of them are .000), while the performance of correlation method 5 and 6 are not significantly different. All correlation methods are slightly better than CombSum in the percentage of runs which are better in "PFB(RR)" than the best component result. Also most of the combination methods (5 out of 6) outperform CombMNZ in this respect.

When comparing correlation method 1 and 2, 3 and 4, and 5 and 6, the experimental results suggest that assigning weights in both ways (on the fly for every query and training by many queries) are equally effective.

Table 5.9 present the performances of these data fusion methods grouping with the same number of results. A common tendency for all these data fusion methods is: their performances increase slightly when we have more results for the fusion (see Figure 5.11 [6]), so do the values of PIB(RR). The values of PFB(RR) are between 54 and 78 in all cases. With 3-7 results, all six correlation methods outperform CombSum and CombMNZ. However, some of the correlation methods do not perform as well as CombSum and CombMNZ with 8 results.

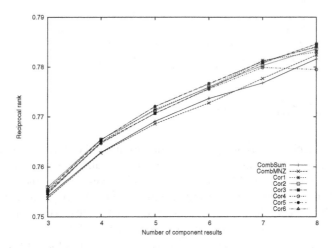

Fig. 5.11 Performance of data fusion algorithms with different number of component results

---

[6] See also Figures 4.5 and 5.5 for a comparison of the same property between CombSum and the linear combination method with two different weighting schemas.

If the correlations among all component results are equal or close to each other, then there is no or very small difference between CombSum and the correlation methods. If the correlations among all component results are quite different, then we can expect that the difference between CombSum and the correlation methods becomes larger. We choose half of all combinations which have stronger correlation then observe the performances of all these data fusion methods. The result is shown in Table 5.10. Compare Table 5.10 with Table 5.8 (for convenience of pairwise comparison, the figures in Table 5.8 is presented in parentheses in Table 5.10), we find that most figures in Table 5.10 are smaller than their counterparts in Table 5.8. This demonstrates that correlation affects data fusion. However, the effect of correlation is not even for different data fusion methods. Because a customized approach has been provided for all correlation data fusion methods, they are able to resist uneven correlation more effectively than CombSum and CombMNZ. In such a situation, all the correlation methods are better than CombSum and CombMNZ at a significance level of .000.

Table 5.10 Summary of performance comparison of data fusion methods (3-8 results, half of 957 combinations having stronger correlation vs all 957 combinations in TREC 6, 7, 8, 9, 2001 and 2002, 50 queries for each combination)

Group	RR	PIB(RR)	PFB(RR)
CombMNZ	0.7662(0.7682)	2.73(3.20)	70.28(73.45)
CombSum	0.7650(0.7689)	2.61(3.34)	68.46(71.82)
Cor1	0.7692(0.7707)	3.09(3.53)	72.90(74.48)
Cor2	0.7706(0.7708)	3.24(3.51)	74.76(76.09)
Cor3	0.7693(0.7704)	3.11(3.54)	73.03(73.86)
Cor4	0.7699(0.7698)	3.19(3.42)	73.79(73.06)
Cor5	0.7711(0.7712)	3.29(3.57)	75.19(75.64)
Cor6	0.7712(0.7712)	3.33(3.60)	75.19(75.69)

5.2.5 Some More Observations

Now based on each year group, we present the results of the above-mentioned experiment. They are shown in Tables 5.11 and 5.12. For all the methods in all year groups, the average value of PFB(RR) is not negative, which shows that these data fusion methods are quite consistent in performance. In Table 5.11, the percentage of runs which are better than the best result varies from about 40 (Cor1, TREC9) to over 90 (several correlation methods in TRECs 6, 8, and 2002).

In all six year groups, 10 systems in TREC 8 were the most accurate, with an average of 0.8196 in RR. Even in such a situation, all data fusion methods perform very well, with an average of between 0.9249 and 0.9308 and all outperforming the best result by over 5%. Moreover, over 95% of the times all data fusion methods outperform the best component result. This suggests that data fusion can be effective with very high accurate retrieval results as long as the other factors are favourable.

Table 5.11 Comparison of six year groups of results (TRECs 6-9 and 2001-2002, performance in RR and PIB(RR))

Method	TREC6	TREC7	TREC8	TREC9	TREC01	TREC02
CombMNZ	0.8886	0.8612	0.9249	0.6962	0.7341	0.5043
	(4.18)	(0.45)	(5.14)	(0.20)	(2.63)	(6.59)
CombSum	0.8820	0.8586	0.9290	0.6964	0.7364	0.5107
	(3.41)	(0.14)	(5.61)	(0.24)	(2.69)	(7.94)
Cor1	0.8908	0.8642	0.9302	0.6960	0.7372	0.5060
	(4.45)	(0.79)	(5.75)	(0.18)	(3.08)	(6.94)
Cor2	0.8903	0.8633	0.9308	0.6983	0.7394	0.5029
	(4.39)	(0.68)	(5.82)	(0.52)	(3.38)	(6.29)
Cor3	0.8886	0.8633	0.9295	0.6965	0.7357	0.5088
	(4.19)	(0.68)	(5.67)	(0.26)	(2.87)	(7.54)
Cor4	0.8866	0.8613	0.9303	0.6977	0.7374	0.5052
	(3.95)	(0.48)	(5.76)	(0.43)	(3.10)	(6.77)
Cor5	0.8924	0.8651	0.9294	0.6954	0.7398	0.5053
	(4.64)	(0.89)	(5.66)	(0.00)	(3.44)	(6.80)
Cor6	0.8890	0.8632	0.9305	0.6980	0.7418	0.5049
	(4.23)	(0.67)	(5.79)	(0.48)	(3.72)	(6.70)

Table 5.12 Comparison of six year groups of results (TRECs 6-9 and 2001-2002, performance in PFB(RR))

Method	TREC6	TREC7	TREC8	TREC9	TREC01	TREC02
CombMNZ	88.61	52.35	95.72	44.97	71.79	87.25
CombSum	82.34	47.23	96.66	42.90	70.64	91.12
Cor1	90.44	55.60	96.68	39.58	74.87	89.72
Cor2	89.86	56.01	97.07	46.31	77.53	89.76
Cor3	88.05	53.81	96.68	40.52	73.11	90.96
Cor4	87.12	50.60	96.78	43.32	73.83	86.68
Cor5	91.90	57.45	96.58	39.59	78.60	89.72
Cor6	90.18	54.54	97.39	40.42	80.36	91.22

We also notice that in two year groups, TREC 7 and TREC 9, all data fusion methods do not perform as well as they do in 4 other year groups on both measures, PIB(RR) and PFB(RR). It is interesting to investigate why this happens.

Let us look at TREC 7 first. In TREC 7, the best result is *CLARIT98CLUS*, whose average performance over 50 queries is 0.8885, 9.4% above the average performance (0.8119) of 9 other results. In such a situation, it is quite difficult for data fusion methods to outperform the best result *CLARIT98CLUS* when it is involved in the fusion. Thus, we divide all combinations into two groups: with and without *CLARIT98CLUS*, which is presented in Table 5.13. The group with *input.CLARIT98CLUS* (Group 1) performs much worse than the group without *CLARIT98CLUS* (Group 2), especially in PFB(RR). In Group 1, all six correlation methods perform better than CombSum, while the difference between CombSum and correlation methods is smaller in Group 2. This is because *CLARIT98CLUS* is very different from the others. The average overlap rate among all possible pairs is 0.4414, while that between *CLARIT98CLUS* and 9 other results is only 0.2780.

Table 5.13 Performance of a group of data fusion methods in TREC 7 (with CLARIT98CLUS - group 1, without CLARIT98CLUS - group 2)

Method	Group 1		Group 2	
	PIB(RR)	PFB(RR)	PIB(RR)	PFB(RR)
CombMNZ	-0.83	26.28	1.80	79.35
CombSum	-1.20	19.72	1.59	76.34
Cor1	-0.47	30.69	2.12	82.58
Cor2	-0.55	29.47	1.97	80.22
Cor3	-0.54	29.47	1.97	80.22
Cor4	-0.74	25.00	1.78	78.28
Cor5	-0.40	32.93	2.26	84.09
Cor6	-0.51	31.10	1.92	79.35

Now let us look at TREC 9. The top two systems are *NEnm* and *NEnmLpas*. Their overlap rate over all 50 queries are 1! Their average performances are 0.7133 and 0.7129, respectively, which are 18.4% and 18.3% above the average performance (0.6027) of 8 other results. Therefore, all data fusion methods do not perform as well as the best component result when both (Group 1 in Table 5.14) or either (Group 2 in Table 5.14) of them are involved in the fusion. When neither of them are involved (Group 3 in Table 5.14), all data fusion methods perform very well.

Table 5.14 Performance of a group of data fusion methods in TREC 9 (Group 1: including top two systems *NEnm* and *NEnmLpas*; Group 2: including either of the top two systems; Group 3: including neither of the top two systems)

Method	Group 1		Group 2		Group 3	
	PIB(RR)	PFB(RR)	PIB(RR)	PFB(RR)	PIB(RR)	PFB(RR)
CombMNZ	-0.75	35.37	-1.66	28.86	5.46	93.15
CombSum	-1.08	30.08	-1.64	25.81	5.92	96.80
Cor1	-1.10	31.71	-1.73	17.89	5.90	98.17
Cor2	-1.06	30.49	-1.70	20.33	6.15	98.17
Cor3	-1.06	30.49	-1.71	20.33	6.15	98.17
Cor4	-0.98	34.15	-1.49	23.98	6.31	98.17
Cor5	-1.40	22.76	-1.67	23.17	5.72	96.23
Cor6	-1.08	28.05	-1.36	21.75	6.37	97.26

Before we finish this section, I would make some comments about CombMNZ. Actually, as we have discussed in Section 4.3, if comparing the score calculating methods of CombSum and CombMNZ, which are $g(d) = \sum_{i=1}^{n} s_i(d)$ (see Equation 1.1) and $g(d) = m * \sum_{i=1}^{n} s_i(d)$ (see Equation 1.2), respectively, they bear some similarity. The difference between them is an extra factor m for CombMNZ. For those documents retrieved by the same number of component systems, their relative ranking positions in the merged results are always the same for both Comb-Sum and CombMNZ. However, when documents are retrieved by different number of component systems, CombSum and CombMNZ will rank them in different ways. Generally speaking, CombMNZ favours those documents which are retrieved by more component systems than CombSum does. We may say that CombMNZ is a special "correlation method" as well. It promotes those documents which are retrieved by more component systems but obtain relatively lower scores while degrading those documents retrieved by fewer component systems but obtain relatively higher scores.

5.3 Combined Weights

In Sections 5.1 and 5.2 we have discussed weighting that considers performance of results and uneven similarity among results, respectively. In this section, we are going to discuss a way of weighting that takes both aspects into consideration. Statistical principles are going to be used. Recall in Section 4.3.1, two conditions, comparable scores and evenly distributed results in the result space, are required for data fusion methods such as CombSum to achieve good results. Now let us discuss the second requirement further to see how to improve the performance of the linear data fusion method. More specifically, the theory of stratified sampling is applied for this [119, 121].

5.3.1 Theory of Stratified Sampling and Its Application

There are two related questions. First, for a group of results (samples), we need to judge if they are uniformly distributed in the sample (result) space or not. Second, if a group of results (samples) are not uniformly distributed, then what is the right way of fusing them to achieve better effectiveness? Both of these questions can be answered using theories of some sophisticated sampling techniques such as stratified sampling [21].

For the population in the sample space, we divide it into m strata S_1, S_2, ..., S_m. Stratum S_h includes N_h units, then there are $N=N_1+N_2+...+N_m$ units in total. In stratum S_h, the value obtained for the i-th unit is denoted as y_{hi}, its weight is $w_h = \frac{N_h}{N}$, and its true mean is $\overline{Y_h} = \frac{\sum_{i=1}^{N_h} y_{hi}}{N_h}$.

Theorem 5.3.1. If in every stratum the sample estimate $\overline{y_h}$ is unbiased, then $\overline{y_{st}} = \frac{1}{N} \sum_{h=1}^{m} N_h \overline{y_h} = \sum_{h=1}^{m} w_h \overline{y_h}$ is an unbiased estimate of the population mean $\overline{Y}$.

Proof: Because the estimates are unbiased in the individual stratum. Therefore, we have

$$E(\overline{y}_{st}) = E\frac{\sum_{h=1}^{m} N_h \overline{y}_h}{N} = \frac{\sum_{h=1}^{m} N_h \overline{Y}_h}{N}$$

But the population mean $\overline{Y}$ may be written as

$$\overline{Y} = \frac{\sum_{h=1}^{m} \sum_{i=1}^{N_h} y_{hi}}{N} = \frac{\sum_{h=1}^{m} N_h \overline{Y}_h}{N}$$

This completes the proof.   □

In stratified sampling, a group of strata are predefined for the population and then the sampling can be processed for every stratum. However, the above process may not be directly applicable to data fusion because the size of each stratum is unknown. What we can do here is to use stratified sampling in the opposite direction. That is to say, for any component result L_i, we try to estimate the size of the stratum that L_i can comfortably represent in order to decide the value of w_i.

If we regard every component result as the mean of a stratum rather than a point in the population space, then according to Theorem 5.3.1, we have $\overline{Y} = \sum_{h=1}^{m} w_h \overline{Y}_h$. This means a weight w_i should be assigned to a result $\overline{Y}_i$ according to the size of the corresponding stratum it represents.

The above discussion demonstrates that data fusion methods such as the correlation methods [119, 121] are reasonable. The remaining question is how to determine the sizes of strata according to what we know about them.

For any two component results (points) $L_1 = \{(d_1, s_{11}), (d_2, s_{12}), ..., (d_n, s_{1n})\}$, and $L_2 = \{(d_1, s_{21}), (d_2, s_{22}), ..., (d_n, s_{2n})\}$, we can calculate the distances $dist(L_1, L_2)$ between them in the result space using Equation 2.15.

For any result L_i, if the distances between it and other results are long, then L_i can be the representative of a large stratum; if the distances are short, then L_i can only be the representative of a small stratum. Note that these strata should be disjoint. Although the information we have is not enough for us to determine the exact sizes of these strata, we can use some heuristic algorithms to estimate the ratios of their sizes. For example, we may define

$$distance(L_i, all) = \frac{1}{n-1} \sum_{k=1 \wedge k \neq i}^{k=n} dist(L_i, L_k)$$

$$distance\_all = \sum_{i=1}^{m} distance(L_i, all)$$

and assign a weight

$$ratio_i = \frac{distance(L_i, all)}{distance\_all} \tag{5.4}$$

to result L_i.

An alternative way of calculating the weights is by using the measure of score dissimilarity. We define the score dissimilarity of two results L_i and L_j as

$$dissimilarity(L_i, L_j) = \sum_{k=1}^{n} |s_{ik} - s_{jk}|$$

then we have

$$diss_i = \frac{1}{n-1} \sum_{k=1 \wedge k \neq i}^{k=n} dissimilarity(L_i, L_k) \tag{5.5}$$

Either $ratio_i$ or $diss_i$ can be used as w_i since both of them possess normalized values between 0 and 1 inclusive.

Training data can be used for the calculation of stratum weights. If this is not available, stratum weights can also be calculated on the fly by using the same data as for fusion.

For each system ir_i, we can define a combined weight, which considers both performance and similarity with others, as

$$w_i = (p_i)^a * (ratio_i)^b \tag{5.6}$$

or

$$w_i = (p_i)^a * (diss_i)^b \tag{5.7}$$

Here a and b are non-negative numbers, p_i is the performance of system ir_i, $ratio_i$ is defined in Equation 5.4, and $diss_i$ is defined in Equation 5.5.

5.3.2 Experiments

The first experiment is conducted with 8 groups of results submitted to TREC [106]. They are results submitted to TRECs 5-8 (ad hoc track), TRECs 9 and 2001 (web track), and TRECs 2003 and 2004 (robust track).

In the experiment, both CombSum (the linear combination) and CombMNZ are involved. For CombMNZ, only the range of [0,1] is used for score normalization; while for CombSum (the linear combination), several different score normalization methods besides [0,1] are used. Firstly, [0.05,1] is used. Here 0.05 is chosen arbitrarily. Secondly, in each year group, we divide all the queries into two groups: odd-numbered queries and even-numbered queries. Then every result is evaluated using AP (average precision) for each query group to obtain performance weight (p_i). The performance weight (p_i) obtained by using the odd-numbered query group is used for fusion of the even-numbered query group, and vice versa. This is called two-way cross validation in [61]. The scores in each result is normalized into range $[p_i/20, p_i]$ to make it comparable with other results, which is in line with the range of [0.05,1] for the normalized scores in the previous option. This method is referred to as the linear combination method with performance weights. Thirdly, all the results are normalized into range [0.05,1] and those documents which are not retrieved in top 1000 were assigned a score of 0, then their strata weights (dis_i or $ratio_i$) are calculated over all available queries according to Equation 5.5 or 5.4. A combined weight w_i for each result is obtained by the product of $diss_i$ and p_i, or the product of $ratio_i$ and p_i. This method is referred to as the combination method with combined weights (performance weights and strata weights).

In each year group, 3-10 results are used for fusion. For a particular number i ($3 \le i \le 10$), 200 runs with i submitted results are randomly selected from all available results in a year group. Then different data fusion methods are used to fuse them and their performances are evaluated by using different measures including average precision (AP), recall-level precision (RP) and precision at certain cut-off document levels (5, 10, 15, 20, 30 and 100). Table 5.15 presents the experimental result in general. Each data point in Table 5.15 is the average of all experimented runs (8 year groups, 8 different numbers (3-10) of component results in each year group, 200 runs for each particular number of component results in each year group) over a group of queries (50 for TREC 5-9 and TREC 2001, 100 for TREC 2003 and 249 for TREC 2004).

Table 5.15 Average performance of different data fusion methods (the figures in parentheses indicate performance improvement of the method over CombSum with the zero-one linear score normalization method)

Measure	CombSum [0,1]	CombMNZ [0,1]	CombSum [0.05,1]	Performance weights	Combined weights
AP	0.2986	0.2975	0.2991	0.3062	0.3072
		(-0.37%)	(+0.17%)	(+2.55%)	(+2.88%)
RP	0.3204	0.3188	0.3207	0.3268	0.3278
		(-0.50%)	(+0.09%)	(+2.00%)	(+2.31%)
P@5	0.5519	0.5524	0.5534	0.5610	0.5636
		(+0.09%)	(+0.27%)	(+1.65%)	(+2.12%)
P@10	0.4874	0.4873	0.4889	0.4958	0.4981
		(-0.00%)	(+0.31%)	(+1.72%)	(+2.20%)
P@15	0.4442	0.4436	0.4455	0.4521	0.4542
		(-0.14%)	(+0.29%)	(+1.78%)	(+2.25%)
P@20	0.4109	0.4102	0.4122	0.4185	0.4203
		(-0.17%)	(+0.32%)	(+1.85%)	(+2.29%)
P@30	0.3622	0.3611	0.3634	0.3691	0.3705
		(-0.30%)	(+0.33%)	(+1.91%)	(+2.29%)
P@100	0.2272	0.2260	0.2278	0.2314	0.2324
		(-0.53%)	(+0.26%)	(+1.85%)	(+2.29%)

From Table 5.15 we can observe that CombSum (with [0.05, 1] score normalization) is slightly better than CombSum (with [0, 1] score normalization) and CombMNZ (with [0, 1] score normalization). The linear combination with performance weights is slightly better than CombSum with [0.05, 1] score normalization, and the linear combination method with combined weights is slightly better than the same method with performance weights. we can observe that the difference between all the methods is not large. Strong correlations exist when using different measures to compare different data fusion methods. The only exception occurs for the comparison of CombSum (with [0,1] normalization) and CombMNZ. When using AP, RP, P@15, P@20, P@30, P@50, and P@100, CombSum performs better than CombMNZ; when using P@5, CombMNZ performs better than CombSum; when using P@10, the performances of CombSum and CombMNZ are almost equal.

The second experiment [111] is carried out with four groups of results submitted to TREC. They are selected runs in the 2001 web track, 2003 and 2004 robust track, and 2005 terabyte track. The zero-one linear normalization method was used for score normalization. The linear combination method with different kinds of combined weights are tested. In $LC(a : b)$, a is the power of p_i and b is the power of $diss_i$ (see Equation 5.7).

Table 5.16 Performance (AP, PIB) of several data fusion methods with combined weights, (In LC(a:b), a and b denote the values of the two parameters in Equation 5.7 for weight calculation.)

Group	LC(1:0)	LC(2:0)	LC(3:0)	LC(3:0.5)	LC(3:1)	LC(3:1.5)	LC(3:2)
2001	0.2637	0.2644	0.2681	0.2686	0.2688	0.2687	0.2682
	(+11.41)	(12.55)	(+13.27)	(+13.48)	(+13.56)	(+13.52)	+(13.31)
2003	0.2865	0.2890	0.2908	0.2912	0.2928	0.2935	0.2938
	(+1.74)	(+2.63)	(+3.27)	(+3.41)	(+3.98)	(+4.23)	(+4.33)
2004	0.3499	0.3522	0.3537	0.3538	0.3535	0.3528	0.3519
	(+5.42)	(+6.12)	(+6.57)	(+6.60)	(+6.51)	(+6.30)	(+6.03)
2005	0.3897	0.3952	0.3986	0.3994	0.4002	0.4009	0.4016
	(+1.94)	(+3.37)	(+4.26)	(+4.47)	(+4.68)	(+4.87)	(+5.05)

Experimental results are shown in Table 5.16. The first three columns present the results in which only performance weights are used. For all other columns, we vary b (0.5, 1, 1.5, and 2) while fixing a to be 3, because 3 appears to be a good value (comparing the first three columns with each other).

Compared with the situation that only performance weights are used, combined weights can bring small but significant improvement (about 0.5%). It suggests that the setting $n = 0.5$ is not as good as the other settings ($b = 1$, 1.5, and 2). However, the three settings ($b = 1$, 1.5, and 2) are very close in performance for data fusion.

Finally, let us compare LC(1:0) (the simple weighting schema) with LC(3.0:1.5) (combined weight, with a performance power of 3 and a dissimilarity power of 1.5). LC(3.0:1.5) performs better than L(1:0) in all 4 year groups by 1% to 3.5%. Although the improvement looks small, it is statistically significant ($p < 0.001$) in all 4 year groups. Since these two schemas require the same relevance judgment information, LC(3.0:1.5) is certainly a better choice than LC(1.0) for data fusion.

5.4   Deciding Weights by Multiple Linear Regression

Suppose there are m queries, n information retrieval systems, and l documents in a document collection D. For each query q^i, all information retrieval systems provide relevance scores to all the documents in the collection, Therefore, we have $(s^i_{1k},$ $s^i_{2k},..., s^i_{nk}, y^i_k)$ for $i = (1, 2, ..., m)$, $k = (1, 2, ..., l)$. Here s^i_{jk} stands for the score assigned by retrieval system ir_j to document d_k for query q^i; y^i_k is the judged relevance score of d_k for query q^i. If binary relevance judgment is used, then it is 1 for relevant documents and 0 otherwise.

Now we want to estimate

$$Y = \{y^i_k; i = (1, 2, ..., m), k = (1, 2, ..., l)\}$$

by a linear combination of scores from all component systems. We use the least squares estimates here, which for the β's are the values $\hat{\beta}_0$, $\hat{\beta}_1$, $\hat{\beta}_2$, ..., and $\hat{\beta}_n$ for which the quantity

$$q = \sum_{i=1}^{m} \sum_{k=1}^{l} [y^i_k - (\hat{\beta}_0 + \hat{\beta}_1 s^i_{1k} + \hat{\beta}_2 s^i_{2k} + ... + \hat{\beta}_n s^i_{nk})]^2$$

is a minimum. This is partly a matter of mathematical convenience and partly due to the fact that many relationships are actually of this form or can be approximated closely by linear equations. β_0, β_1, β_2,..., and β_n, the multiple linear regression coefficients, are numerical constants that can be determined from observed data.

In the least squares sense the coefficients obtained by multiple linear regression can bring us the optimum fusion results by the linear combination method, since they can be used to make the most accurate estimation of the relevance scores of all the documents to all the queries as a whole. In a sense this technique for the improvement of fusion performance is general since it is not directly associated with any ranking-based metric such as average precision or recall-level precision. Theoretically, this approach works perfectly with score-based metrics such as the Euclidean distance (see Section 6.4), but it should work well with any reasonably defined ranking-based metrics such as average precision or recall-level precision as well, because more accurately estimated scores for all the documents are able for us to obtain better rankings of those documents.

In the linear combination method, those multiple linear regression coefficients, β_1, β_2,..., and β_n, can be used as weights for ir_1, ir_2,..., ir_n, respectively. We do not need to use β_0 to calculate scores, since the relative ranking positions of all the documents are not affected if a constant β_0 is added to all the scores.

Let us take an example to illustrate how to obtain the multiple regression coefficients from some data. Suppose we have a collection of 4 documents, 3 information retrieval systems, and 2 queries. For each query, each information retrieval system assigns a score to each document in the collection. Binary relevance judgment is used. The scoring information and relevance judgment of all the documents are shown below.

Query	Doc	ir_1	ir_2	ir_3	Relevance
q_1	d_1	.50	.30	.80	1
q_1	d_2	.60	.70	.40	1
q_1	d_3	.10	.80	.40	0
q_1	d_4	.20	.30	.10	0
q_2	d_1	.30	.40	.80	1
q_2	d_2	.20	.50	.10	0
q_2	d_3	.30	.40	.40	0
q_2	d_4	.30	.50	.50	1

By setting ir_1, ir_2, and ir_3 as independent variables, and *relevance* as the dependent variable, we use R[7] with the "Rcmdr" package to obtain the multiple linear regression coefficients. In this example, we obtain $\beta_1=0.498$, $\beta_2=0.061$, and $\beta_3=1.979$ for ir_1, ir_2, and ir_3, respectively.

5.4.1 Empirical Investigation Setting

In this subsection let us see the experimental settings and some related information with five groups of runs submitted to TREC. They are selected runs submitted to TRECs 5-8 ad hoc track and TREC 9 web track. The results will be discussed in the following three subsections.

The binary logistic model (see Section 3.3.4) is used for generating normalized scores needed. The parameters we obtain are shown in Table 5.17.

Table 5.17 Coefficients obtained using binary logistic regression for five year groups TRECs 5-9

Group	Odd-numbered group		Even-numbered group		All-together	
	a	b	a	b	a	b
TREC 5	.904	-.670	.762	-.727	0.821	-0.692
TREC 6	1.048	-.722	.594	-.729	0.823	-0.721
TREC 7	1.209	-.764	1.406	-.771	1.218	-0.756
TREC 8	1.218	-.796	1.538	-.750	1.362	-0.763
TREC 9	.448	-.757	.368	-.690	-	-

Note that in a year group, for any query and any result (run), the score assigned to any document is only decided by the ranking position of the document. For example, if $a=1.218$ and $b=-.765$. It follows from Equation 3.10 that

$$score(t) = \frac{1}{1 + e^{-1.218 + 0.765 * ln(t)}} \qquad (5.8)$$

---

[7] R is a free statistical software. Its web site is located at http://www.r-project.org/

Therefore, $score(1) = 0.7717$, $score(2) = 0.6655$,..., and so on. With such normalized scores for the runs selected, the next step is to use multiple linear regression to train system weights and carry out fusion experiments. For all 50 queries in each year group, we divide them into two groups: odd-numbered queries and even-numbered queries. As before, we use the two-way cross validation.

The file for multiple linear regression is a m row by $(n + 1)$ column table. Here m is the total number of documents in all the results and n is the total number of component systems involved in the fusion process. In the table, each record i is used for one document d_i ($1 \leq i \leq m$) and any element s_{ij} ($1 \leq i \leq m$, $1 \leq j \leq n$) in record i represents the score that document d_i obtains from component system ir_j for a given query. Any element $s_{i(n+1)}$ is the judged score of d_i for a given query. That is, 1 for relevant documents and 0 for irrelevant documents. The file can be generated by the following steps:

1. For each query, go through Steps 2-5.
2. If document d_i occurs in at least one of the component results, then the related information of d_i is put into the table as a row. See Steps 3-4 for more details.
3. If document d_i occurs in result r_j, then the score s_{ij} is calculated out by the logistic model (parameters are shown in Table 5.17).
4. If document d_i does not occur in result r_j, then a score of 0 is assigned to s_{ij}.
5. Ignore all those documents that are not retrieved by any component systems.

The above Step 5 implies that any document that is not retrieved will be assigned a score of 0. In this way we can generate the table for multiple linear regression.

5.4.2 Experiment 1

We first investigate the situation with various different number of component systems, from 3, 4, ..., to up to 32 component systems [108]. For each given number k ($3 \leq k \leq 32$), 200 combinations are randomly selected and tested. Eight metrics, including average precision, recall-level precision, precision at 5, 10, 15, 20, 30, and 100 document level, are used for performance evaluation.

Apart from the linear combination method with the trained weights by multiple regression (referred to as LCR), CombSum [32, 33], CombMNZ [32, 33], Pos-Fuse [51], MAPFuse [51], SegFuse [82], the linear combination method with performance level weighting (referred to as LCP [9, 94]), and the linear combination method with performance square weighting (referred to as LCP2, [112]) are also involved in the experiment.

Scores can be converted from ranking information with some training data in different ways. In MAPFuse [51], a document at rank t is assigned a score of v_{map}/t, where v_{map} is the MAP (mean average precision) value of the system in question over a group of training queries using a training document collection. PosFuse [51] also uses a group of training queries and a training document collection. For the ranked list of documents from each component system, the posterior probabilities of documents at each given rank are observed and averaged over a group of queries. Then the averaged posterior probability is used as relevance scores for the document at that rank. In SegFuse [82], scores are generated using a mixture of normalized raw scores (s_{n-rs}, raw scores are provided by the information retrieval system) and ranking-related relevance scores (s_p, from ranking-related posterior probabilities). For s_{n-rs}, the zero-one linear score normalization method is used to normalize them. For obtaining s_p, SegFuse also needs a group of training queries and a document collection. However, it takes a slightly different approach from PosFuse. Instead of calculating posterior probabilities of documents at each given rank, SegFuse divides the whole lists of documents into segments of different size ($size_k = (10 * 2^{k-1} - 5)$). In each segment, documents are assigned equal relevance scores. The final score of a document is given by $s_p(1 + s_{n-rs})$. For the fusion process, the above three methods just use the same method as CombSum does.

Figures 5.12-5.14 present the results.

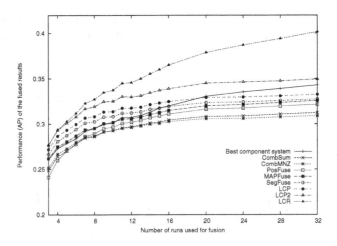

Fig. 5.12 Performance comparison of different data fusion methods (average of 5 groups and 200 combinations for each given number of component systems, AP)

From Figure 5.12, we can see that LCR is the best in all year groups. If we consider the average of all the combinations (5 year groups * 18 different numbers of component systems * 200 combinations for each given number of component systems * 50 queries), then LCR outperforms LCP2, LCP, SegFuse, MAPFuse, PosFuse, CombSum, CombMNZ, and the best component systems by 5.81%, 9.83%, 11.75%, 13.70%, 16.27%, 17.79%, 17.96%, and 12.35%, respectively. Two-tailed t test (see Table 5.18) shows that the difference between LCR and the others are very highly significant (p value < 0.01).

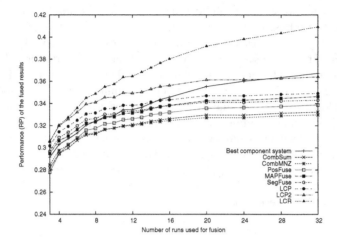

Fig. 5.13 Performance comparison of different data fusion methods (average of 5 groups and 200 combinations for each given number of component systems, RP)

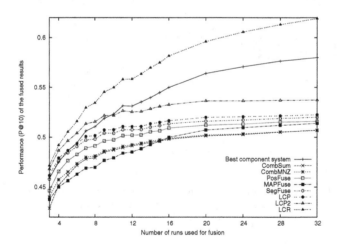

Fig. 5.14 Performance comparison of different data fusion methods (average of 5 groups and 200 combinations for each given number of component systems, P@10)

Table 5.18 Results of two-tailed t test of comparing LCR and three other data fusion methods (200 randomly selected combinations for 3-16, 20, 24, 28, and 32 component systems; degrees of freedom are 19999; in all cases, p-value<2.2e-16)

Pair of methods	t value	Mean of the differences
LCR & SegFuse	81.57	0.0363
LCR & LCP	71.86	0.0306
LCR & LCP2	54.70	0.0188
LCR & the best system	119.90	0.0377

Considering the difficulty of beating the best component systems, an average improvement of 12.35% is a very good result. More importantly, such an improvement is consistent across all metrics and all year groups. On average, LCP and LCP2 also outperform the best component system by 2.33% and 6.25%, respectively. This confirms that LCP and especially LCP2 are also effective data fusion methods. However, they are not as consistent as LCR when different numbers of systems are fused or different metrics are used for evaluation. When 3 or 4 component systems are fused, LCP2 is almost as good as LCR. However, when more than 4 component systems are fused, LCP2 is not as good as LCR and the difference between LCP2 and LCR becomes greater and greater. SegFuse is marginally better than the best component system (by 0.56%), while the other four methods MAPFuse, PosFuse, CombSum, and CombMNZ are not as good as the best component system.

When using the metric of recall-level precision (see Figure 5.13), the improvement rates of LCR over other methods are: 5.03% (LCP2), 8.22% (LCP), 8.75% (the best system), 10.02% (SegFuse), 10.34% (MAPFuse), 12.80% (PosFuse), 14.40% (CombSum), and 14.51% (CombMNZ). On average, LCP and LCP2 perform slightly better than the best component system by 1.80% and 4.77%, respectively. However, LCP is not as good as the best component system when 15 or more systems are fused, LCP2 is not as good as the best component system when 28 or 32 systems are fused. All other fusion methods are not as good as the best component system on average.

If considering precision at 10 document level (see Figure 5.14), then the improvement rate of LCR over other methods are: 6.88% (LCP2), 9.76% (LCP), 5.10% (the best system), 10.38% (SegFuse), 14.85% (MAPFuse), 11.70% (PosFuse), 14.46% (CombSum), and 13.99% (CombMNZ). Apart from LCR, all other fusion methods are worse than the best component system on average, though LCP and LCP2 perform a little better than the best component system when a few component systems are fused.

In order for us to have a better understanding of the properties of all the data fusion methods involved, we investigate the relation of data fusion methods to the best component system and to all component systems. We calculate Pearson product-moment correlation coefficients of performances between data fusion methods and the best component system, and coefficients between data fusion methods and the average of all component systems. A good data fusion method should have a strong positive correlation with the best component system rather than the average of all component systems. The result of this experiment is shown in Tables 5.19 and 5.20. These coefficients vary from one year group to another even for the same data fusion method. However, in all year groups, LCR has the biggest Pearson's coefficients with the best component system, while it has the smallest Pearson's coefficients with the average of all component systems. This confirms that LCR is the best fusion method from a different angle.

Table 5.19 Pearson's correlation coefficient of performances (AP) between data fusion methods and the best component system

Methods	Average	Maximum	Minimum
CombSum	0.7376	0.7846	0.6870
CombMNZ	0.7164	0.7430	0.6568
PosFuse	0.7922	0.8553	0.7259
MAPFuse	0.8624	0.9225	0.8031
SegFuse	0.8353	0.8875	0.7616
LCP	0.8523	0.9091	0.7809
LCP2	0.8293	0.9639	0.8502
LCR	0.8947	0.9738	0.8468

Table 5.20 Pearson's correlation coefficient of performances (AP) between data fusion methods and the average of all component systems

Methods	Average	Maximum	Minimum
CombSum	0.6782	0.7233	0.6172
CombMNZ	0.6593	0.7076	0.5770
PosFuse	0.4763	0.5513	0.4154
MAPFuse	0.5627	0.6652	0.4949
SegFuse	0.5398	0.6659	0.4463
LCP	0.5194	0.6573	0.4335
LCP2	0.4936	0.5962	0.3959
LCR	0.4114	0.5244	0.3261

By looking at the experimental results evaluated by different metrics (average precision, recall-level precision, precision at 10 document level, etc.) at the same time, we have a few further observations.

1. We can see that LCR consistently outperforms all other methods and the best component system. When the number of component systems increases, the difference in performance between LCR and all other data fusion methods increases accordingly. When 32 systems are fused, LCR outperforms other data fusion methods by very large margins:

Pair of methods	AP	RP	P@10
LCR & LCP2	14.93%	12.45%	15.28%
LCR & LCP	20.78%	17.25%	18.44%
LCR & SegFuse	22.68%	19.28%	19.01%
LCR & MAPFuse	23.25%	18.15%	20.46%
LCR & PosFuse	24.92%	20.80%	19.95%
LCR & CombSum	28.58%	23.15%	22.06%
LCR & CombMNZ	29.97%	24.10%	22.15%
LCR & the best component system	17.04%	11.40%	6.71%

2. LCP2 is always better than LCP, which confirms that performance square weighting is better than performance level weighting.

3. Compared with the best component system, LCP and LCP2 perform better when a small number of component systems are fused. However, there are significant differences when different metrics are used. Average precision is the metric that favours LCP and LCP2, while precision at 10 document level favours the best component system.

4. CombSum and CombMNZ perform badly compared with all the linear combination data fusion methods and the best component system. The result is understandable since no training is needed for both of them. It also demonstrates that treating all component systems equally is not a good fusion policy when there are a few poorly performing component systems.

5. As to effectiveness, the experimental results demonstrate that we can benefit from fusing a large group of component systems no matter which fusion method we use. Generally speaking, the more component systems we fuse, the better the fusion result we can expect. This can be seen from all those curves that are increasing with the number of component systems in all three figures 5.12-5.14 in this subsection.

6. It suggests that LCR can cope well with all sorts of component systems that are very different in effectiveness, even some of the component systems are very poor. It can bring effectiveness improvements steadily for different groups of component systems no matter which rank-based metric is used.

5.4.3 Experiment 2

Compared with some numerical optimization methods such as conjugate gradient, the multiple linear regression technique is much more efficient and much more likely to be used in real applications. However, if a large number of documents and a large number of information retrieval systems are involved, then efficiency might still be an issue that needs to be considered. For example, in some dynamic situations such as the web, documents are updated frequently. Weights may need to be calculated quite frequently so as to obtain good performance. Therefore, we investigate if it is possible for us to use the linear regression technique in a more efficient way.

An experiment is carried out to compare the performance of the linear combination data fusion method with two groups of different weights [110]. One group uses

the training data set of all 1000 documents; the other group uses the training data
set that only includes top 20 documents for each of the queries. All other aspects
of the methodology are just the same as in Section 5.4.2. Four groups of results
(TRECs 5-8) are used. The experimental result is shown in Figure 5.15. In Figure 5.15, LCR_1000 represents the LCR method using the 1000-document training
set and LCR_20 represents the LCR method using the 20-document training set. In
Figure 5.15, for all three metrics AP, RP, and P@10, the curves of LCR_1000 and
LCR_20 overlap with each other and are totally indistinguishable. In almost all the
cases, the difference between them is less than 0.3%. this demonstrates that using
the top-20 documents for each of the queries as the training data set can be equally
effective as using all 1000 documents. Obviously, using fewer documents in the
training data set is beneficial if efficiency is a major concern.

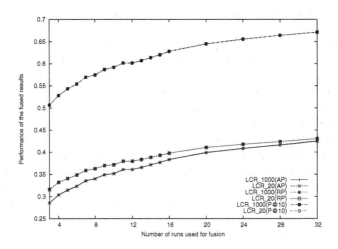

Fig. 5.15 Performance comparison of the linear combination fusion methods with two groups
of different weights trained using different data sets (average of 4 groups and 200 combinations for each given number of component systems)

We also measure the time needed for weighting calculation by multiple linear regression. A personal computer[8], which installed R with Windows interface package
"Rcmdr", and two groups of data, TREC 5 and TREC 7 (even-numbered queries)
were used for this part of the experiment. The time needed for the multiple linear
regression is shown in Table 5.21. Note that in this experiment, for a year group, we
imported data of all component systems into the table and randomly selected 5, 10,
15, or 20 from them to carry out the experiment. It is probably because the TREC 7

---

[8] It is installed with Intel Duo core E8500 3.16GHz CPU, 2 GB of RAM and Windows
operating system.

group comprises more component systems (82) than the TREC 5 group does (53), it took a little more time for the multiple regression to deal with TREC 7 than TREC 5 when the same number of component systems (5, 10, 15, or 20) are processed. We can see that roughly the time needed for the 20-document data set is 1/20 of the time for the 1000-document data set, though the number of records in the 20-document data set is 1/50 of that in the 1000-document data set. But a difference like this may still be a considerable advantage for many applications. It is also possible to speed up the process if we only upload those data needed into the table.

Table 5.21 Time (in seconds) needed for each multiple linear regression run

Number of variables	TREC 5		TREC 7	
	1000	20	1000	20
5	1.20	0.06	1.40	0.07
10	1.60	0.08	1.70	0.09
15	2.00	0.10	2.80	0.12
20	2.70	0.14	3.60	0.14

5.4.4 Experiment 3

In this subsection we investigate the effect of different component systems, especially, some very poor component systems, on the performance of the linear combination data fusion method. As we know, if considering data fusion methods such as CombSum and CombMNZ, very poor component systems will affect the performance of fusion significantly; therefore, they should be removed by all possible means. For the linear combination method, it may be different from CombSum and CombMNZ, since different weights can be assigned to different systems. We hypothesize that even very poor component systems should not be harmful, if they cannot help improve the performance of data fusion at all. It is interesting to find out what the truth is.

We carry out an experiment as such: we randomly select $3i$ ($i = 1, 2,..., 10$) systems from all possible ones in a year group, then we remove one third of the poorest component systems (measured by AP), thus we have $2i$ ($i = 1, 2,..., 10$) component systems; then we remove one half of the poorest component systems from the remaining ones (measured by AP), and we have the best i ($i = 1, 2,..., 10$) component systems. Thus we obtain three groups of related component systems and each of these three groups of component systems are fused by LCR in the same way as in Experiments 1 and 2, then we can compare their performances so as to have some findings. Four groups of results (TRECs 5-8) are used. Figures 5.16- 5.18 present the results.

From Figures 5.16-5.18, we can see that the group of all systems does slightly better than the group of top two third systems, and the group of top two third does

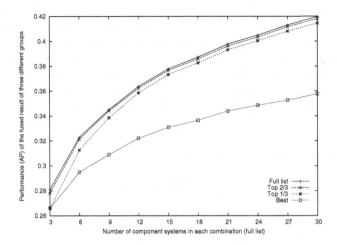

Fig. 5.16 Performance comparison of the linear combination fusion methods with three groups of component systems (full list, top one third of the full list, top two third of the full list, average of 4 groups and 200 combinations for each given number of component systems, average precision)

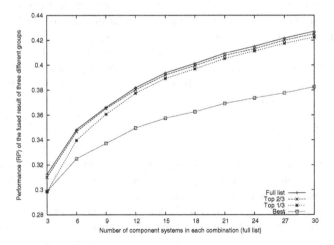

Fig. 5.17 Performance comparison of the linear combination fusion methods with three groups of component systems (full list, top one third of the full list, top two third of the full list, average of 4 groups and 200 combinations for each given number of component systems, precision-level recall)

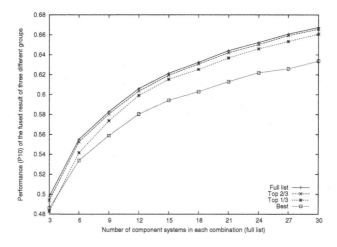

Fig. 5.18 Performance comparison of the linear combination fusion methods with three groups of component systems (full list, top one third of the full list, top two third of the full list, average of 4 groups and 200 combinations for each given number of component systems, precision at 10 document level)

better than the group of top one third on all measures including AP, RP, P@10. Taking AP as an example, on average, the full-list is better than the top two-third by 0.45%, the top two-third is better than the top one-third by 1.48%, and the full-list is better than the top one-third by 1.95%. All the differences for 3(2,1), 6(4,2), 9(6,3), ..., 30(20,10) component systems are highly significant (p-value $< 2.2e\text{-}16$ in pairwise two-tailed t test). For RP, P@10 and other metrics, the corresponding percentages of improvement are very similar. From this experiment, we can see that even very poor component results are not just harmful, but a little helpful, to the performance improvement of the fused results. On the other hand, Figure 5.19 shows the performance (average precision) of CombSum and CombMNZ on the same component systems. It can be seen that there is considerable difference between these three groups of systems. On average, the top one-third group is better than the top two-third group by 6.14%, the top two-third group is better than the full-list group by 5.52%, and the top one-third group is better than the full-list group by 11.89%. The top one-third group is better than the best component system by a clear margin, the top two-third group is sometimes better than the best component system but sometimes worse than the best component system, while the full-list group is always worse than the best component system.

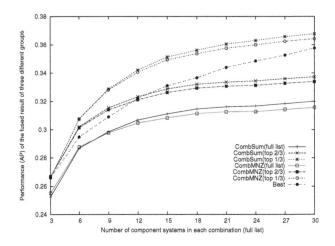

Fig. 5.19 Performance comparison of CombSum and CombMNZ with three groups of component systems (full list, top one third of the full list, top two third of the full list, average of 4 groups and 200 combinations for each given number of component systems, average precision)

5.5 Optimization Methods

Some sophisticated optimization methods might be used for this purpose in many different ways. Two examples are:

Bartell and Cottrell [9] used conjugate gradient, a numerical optimization method, to find good weights for component systems. Conjugate gradient is an iterative optimization method for finding the nearest local maximum of a function of n variables which presupposes that the gradient of the function can be computed. The algorithm is computation-intensive and time-consuming. Due to that, only 2 to 3 component systems and top 15 documents returned from each system for a given query were used in their investigation.

Vogt and Cottrell [98, 99] analysed the performance of the linear combination method. In their experiments, they used all possible pairs of 61 systems submitted to the TREC 5 ad-hoc track. Another numerical optimization method, the golden section search, was used to search for best system weights. Due to the nature of the golden section search, only two component systems can be involved for each fusion case.

In my personal point of view, to use iterative optimization methods such as conjugate gradient is less useful for such a task in practical applications. However, in some special situations, they might be helpful. One situation is that the interest of the users is focused on one particular measure (e.g., reciprocal rank, precision at 5

Table 5.22 Performance (measured by AP) of the fused result using the linear combination method, in which suitable weights are obtained by the conjugate gradient method (TREC 2001 web track, 3 component results, 200 runs)

Starting weights	AP (starting weights)	AP (searched weights)	Improvement of (3) over (2)	Time (Seconds)
0.5	0.23931	0.23945	+0.06%	42.8
p	0.23976	0.23984	+0.03%	30.5
p^2	0.24176	0.24193*	+0.07%	47.1
p^3	0.24252	0.24402*	+0.62%	88.3
p^4	0.24280	0.24316*	+0.15%	58.4
p^5	0.24200	0.24200	+0.0%	16.5

document level). Then optimization methods might be a good option. Another situation is that there is a very strong demand for excellent weighting performance. we might be able to obtain better results if using two methods together and one of them is an optimization method.

Now let us see how the power weighting schema, which we discussed in Section 5.1, can be used with optimization methods together. For a set of component systems over a group of queries, we would like to find better weights for all of them. By using the power weighting schema, we assign a weight to each retrieval system involved. Then those weights are used as the starting point for the optimization method. Such a preparation is reasonable because we consider that the global maximum could be more easily reachable in a fewer steps. It is good for both effectiveness and efficiency.

An experiment [112] is carried out with 200 fusion runs, and each of them includes 3 component results which were chosen randomly from the runs submitted to the TREC 2001 web track [9]. As in [9], we used the conjugate gradient method. For each component result involved, we evaluate its average performance (in AP) over all 50 queries. Suppose that the three component results are L_1, L_2, and L_3, their average performances are p_1, p_2, and p_3, respectively. Then we use p_i ($i = 1$, 2, and 3) as L_i's initial weight to begin the search process. In addition, several other options p_i^2, p_i^3, p_i^4, p_i^5, and 0.5 (equal weight for all component results) are also used as initial weights. It seems that there are multiple local maxima in the search space, because different starting weights usually lead to different maxima (local maxima). The conjugate gradient method, like many other optimization methods, is always able to find a local maximum, but not always a global maximum. Table 5.22 shows the experimental result.

Note that the starting point can affect both effectiveness and efficiency of optimization methods. Firstly, if the starting point chosen is very close to a local or global maximum, then it takes only a few steps for the optimization method to approach it. Otherwise, it may take much longer. Thus the efficiency of the optimization method can be affected. Secondly, different starting points lead to different

---

[9] Only those runs whose average precision was over 0.15 were used.

(local or global) maxima. It is very likely that for any starting point, the closest local or global maximum will be found. Therefore, the effectiveness of the optimization method can be affected by the starting point as well. In Table 5.22, p^3 leads to the best results among all the options. Therefore, a proper power weighting schema can be used as a good starting point for the optimization method to search for more favourable weights. Comparing the figures in column 3 with the figures in column 2, some of them, with a "*" mark, are statistically significant (p-value ≤ 0.05).

The time required for each run of the conjugate gradient method is given in Table 5.22 as well. An Intel Dual-core CPUs and 1GM of RAM PC was used for this. At the time that the experiment was conducted (in 2008), this PC is reasonably good. For each group of 3 component results over 50 queries, the time required varies from 16 seconds to 1.5 minutes. However, in the case of 16 seconds for p^5, no improvement has been made. Since the optimization method cannot find any better points in the neighbourhood and stops the search process after only very few steps. The more time the search process takes, the larger improvement on fusion we can obtain (comparing the best point found with the starting point). More time is needed if more component results or more queries are involved. Especially when more component results are involved, the time complexity of the search process increases very rapidly.

5.6 Summary

In Sections 5.1-5.3, a few different heuristic methods of assigning appropriate weights for the linear combination methods have been discussed. The methods discussed in Section 5.1 only concern performance of component systems, the methods discussed in Section 5.2 only concern uneven similarity among different component systems, while the methods discussed in Section 5.3 concern both performances of component systems and uneven similarity among different component systems at the same time. In Section 5.4, multiple linear regression is used to find the best suitable weights in the sense of least squares. Section 5.5 addresses iterative optimization methods. We may use them to define an objective, so as to maximize the performance of the data fusion method measured by a given metric, which is difficult to express accurately by more efficient methods or use them together with other heuristic methods for obtaining more profitable weights.

In all these weighting schemas discussed so far, which one is the best for us to use? It depends on the situation. Most of the time, multiple linear regression is a very effective technique for weights assignment. However, it requires more information than some other methods. On the other hand, for the performance-related weighting, we only need very little information. For example, This can be by using only a few training queries and the results are evaluated by RR, P@5, or P@10. The weights obtained in such a way can still be reasonably good. If all component systems are very close in performance, but uneven similarity among them is the major concern, then the correlation methods can be a good option.

Chapter 6
A Geometric Framework for Data Fusion

Quite a few data fusion methods have been proposed, but questions such as why data fusion can bring improvement in effectiveness and what are the favourable conditions for data fusion algorithms are only partially or vaguely answered due to the uncertainty of the problem. In this chapter, we set up a geometric framework to formally describe score-based data fusion methods, in which each component result returned from an information retrieval system for a given query is represented as a point in a multi-dimensional space. The performance of any result and the similarity between any pair of results can be evaluated by the same metric – the Euclidean distance. Then all the component results and the fused results can be explained using geometrical principles. In such a framework, data fusion becomes a deterministic problem. The performance of the fused result is determined by the performances of all component results and the similarities among all of them. Several interesting features of the centroid-based data fusion method and the linear combination method can be deduced. As a formal model of data fusion, this framework enable us to have a better understanding of the nature of data fusion and use the data fusion technique more precisely and effectively [105].

6.1 Relevance Score-Based Data Fusion Methods and Ranking-Based Metrics

Quite a few data fusion methods, such as CombSum [32, 33], CombMNZ [32, 33], the linear combination method [98, 99, 112, 113], are score-based methods. For score-based methods, all component retrieval systems need to provide relevance scores for all retrieved documents. Those scores are calculated in one way or another to produce the fused result. On the other hand, almost all commonly used measures, such as average precision (AP), recall-level precision (RP), precision at certain cut-off document level (P@k), and many others, are ranking-based measures. For those ranking-based measures, only the ranking positions of relevant and irrelevant documents matter. Many ranking-based measures have been used for many years and

S. Wu: Data Fusion in Information Retrieval, ALO 13, pp. 117–133.
springerlink.com   © Springer-Verlag Berlin Heidelberg 2012

they should be fine for retrieval evaluation. However, when using them to investigate the data fusion problem, especially those score-based data fusion methods, it introduces uncertainty to the problem: compared with component results, the fused result may go either direction: better or worse in effectiveness. Let us see some examples to illustrate this.

Example 6.1.1. Suppose we have a collection of documents and two information retrieval systems. For a given query q, there is only 1 relevant document (d_2) in the whole collection. One information retrieval system returns a ranked list of 3 documents $L_1 = <(d_1, 0.7), (d_2, 0.5), (d_3, 0.1)>$, and the other information retrieval system also returns a ranked list of 3 documents $L_2 = <(d_3, 0.6), (d_2, 0.5), (d_1, 0.1)>$.

We can fuse these two results by averaging the scores of the documents involved and the fused result $F = <(d_2, 0.5), (d_1, 0.4), (d_3, 0.35)>$. In this case, data fusion is successfully promoting the relevant document to the front of all irrelevant documents, and both AP and RP values of the fused result are much better than that of the two component results. If using AP and RP to evaluate them, then we have $AP(L_1) = 1/2$, $RP(L_1) = 0$, $AP(L_2) = 1/2$, $RP(L_2) = 0$, $AP(F) = 1$, $RP(F) = 1$. □

Example 6.1.2. Suppose there are two relevant documents (d_1 and d_2) in the whole collection. One component result $L_1 = <(d_1, 0.7), (d_3, 0.6), (d_2, 0.4)>$, the other result $L_2 = <(d_2, 0.7), (d_3, 0.6), (d_1, 0.3)>$, and the fused result $F = <(d_3, 0.6), (d_2, 0.55), (d_1, 0.5)>$ is obtained by averaging all the scores.

In this case, both component results have relevant documents at ranking positions 1 and 3. However, the fused result has the relevant documents at ranking positions 2 and 3. Therefore, the fused result is worse than both component results if measured by AP, RP, or some other ranking-based measures. □

Example 6.1.3. Suppose there are 3 relevant documents (d_1, d_4, d_5) in the whole collection, two component results $L_1 = <(d_2, 0.90), (d_1, 0.70), (d_4, 0.50), (d_5, 0.40), (d_3, 0.30)>$, $L_2 = <(d_3, 0.80), (d_1, 0.70), (d_4, 0.50), (d_5, 0.40), (d_2, 0.30)>$, using the averaging method, the fused result is $F = <(d_1, 0.70), (d_2, 0.60), (d_3, 0.55), (d_4, 0.50), (d_5, 0.40)>$.

If we evaluate them using AP, then $AP(L_1) = AP(L_2) = 1/3(1/2+2/3+3/4) = 23/36$, $AP(F) = 1/3(1/1+2/4+3/5) = 32/60$, thus $AP(L_1) = AP(L_2) > AP(F)$.

If we use RP to evaluate them, then $RP(L_1) = RP(L_2) = 2/3$, $RP(F) = 1/3$, thus $RP(L_1) = RP(L_2) > RP(F)$.

If we use RR to evaluate them, then $RR(L_1) = RR(L_2) = 1/2$, $RR(F) = 1$, thus $RR(L_1) = RR(L_2) < RR(F)$. □

Examples 6.1.1 and 6.1.2 show that using those ranking-based measures, the performances of the fused results by data fusion are mixed if compared with that of component results. Further, Example 6.1.3 shows that for the same result, we may obtain totally different conclusions if using different measures to evaluate them. If we try to understand what is going on from one case to another, there will be no certain solution. We may use statistical methods to analyse some experimental results. The difficulty is that the observations and conclusions may vary when different data sets are used. Therefore, that is why some observations were made, but they have been disapproved later on. Some hypotheses remain to be hypotheses since they can neither be verified nor disapproved. However, we find that uncertainty is not necessarily a characteristic of relevance score-based data fusion methods. If we also use relevance score-based measures such as the Euclidean distance to evaluate retrieval results, then the data fusion problem can be a deterministic problem. That means, the conclusions we make will hold in every single case. This is certainly very helpful for us to have a better understanding of the nature of the data fusion problem and make the most profit from data fusion in information retrieval.

6.2   The Geometric Framework

Suppose we have a collection of p documents $D=(d_1, d_2,..., d_p)$ and a group of m information retrieval systems $IR=(ir_1, ir_2,... , ir_m)$. For a given query q, a retrieval system returns a set of scores $S_j = (s_j^1, s_j^2,...., s_j^p)$ for all the documents as the result. Here each s_j^k ($1 \leq j \leq m$, $1 \leq k \leq p$) denotes the estimated relevance degree of document d_k to q by ir_j. The range of any s_j^k is between 0 and 1 inclusive. For example, a relevance degree of 0.6 denotes that the document is 60% relevant to the query. This framework is very general since different kinds of relevance judgment can be supported. If l ($l \geq 2$) graded relevance judgment is used, then we can assign a relevance score $(k-1)/(l-1)$ for the documents at grade k ($0 \leq k \leq l-1$). For example, if $l = 5$, then the scores $\{0, 0.25, 0.5, 0.75, 1\}$ are assigned to documents at grades $\{0, 1, 2, 3, 4\}$, respectively.

If all retrieved documents are assigned relevance scores, then we can define a p-dimensional space, in which the relevance degree of each document is regarded as one dimension, then any retrieval result $S_j = (s_j^1, s_j^2,..., s_j^p)$ must be located inside a hypercube of p dimensions. The hypercube is referred to as the result space for a collection of documents and a given query. It is denoted as $X(q, D)$ or X in short, where q is the query and D is the document collection. The dimension of the space is the number of documents in D. In $X(q, D)$, a result is equivalent to a point.

Definition 6.2.1. (Euclidean distance[1]) $S_1 = (s_1^1, s_1^2,..., s_1^p)$ and $S_2 = (s_2^1, s_2^2,..., s_2^p)$ are two points in $X(q, D)$, the Euclidean distance between S_1 and S_2 is defined as

$$dist(S_1, S_2) = \sqrt{\sum_{k=1}^{p} (s_1^k - s_2^k)^2} \qquad (6.1)$$

In $X(q, D)$, one point $O = (o^1, o^2,..., o^p)$ is the ideal result, in which every o^k is the real relevance score of the corresponding document d^k. For any result $S = (s^1, s^2,..., s^p)$, its performance can be evaluated by calculating the Euclidean distance $(dist)$ of this result from the ideal point in the result space (i.e., $dist(S, O) = \sqrt{\sum_{k=1}^{p} (s^k - o^k)^2}$). See also Section 2.4 for more details of score-based metrics.

---

[1] It was defined in Equation 2.15. For convenience, we give its definition here again.

In $X(q, D)$, data fusion can be described as finding a result F, which is fused from a group of results $S_j = (s_j^1, s_j^2, ..., s_j^p)$ $(1 \leq j \leq m)$ in some way. We expect that F can be as close to the ideal point as possible.

The Euclidean distance can also be used to measure the similarity of two component results, which is one of the factors that affect the effectiveness of data fusion significantly [64, 121, 123]. In this geometrical framework, both result effectiveness and similarity between results can be measured by using the same measure – the Euclidean distance, which is a big advantage for us to further investigate the features of some major data fusion methods.

6.3 The Centroid-Based Data Fusion Method

For a group of results $S_j = (s_j^1, s_j^2, ..., s_j^p)$ $(1 \leq j \leq m)$, there is a point C that is the centroid of these points $S_1, S_2, ..., S_m$. Set $C = (c_1, c_2, ..., c_p)$, then

$$c^k = \frac{1}{m} \sum_{j=1}^{m} s_j^k \tag{6.2}$$

for any k $(1 \leq k \leq p)$. One major data fusion method is to average scores or to calculate the centroid of a group of component results as the fused result, which is referred to as the centroid-based data fusion method. Note that the centroid-based method is the same as CombSum [32, 33] if we only concern the ranking of documents in the fused results or use ranking-based measures to evaluate them. Next we discuss some interesting properties of the centroid-based method.

Theorem 6.3.1. In a p-dimensional space X, there are m points $S_j = (s_j^1, s_j^2, ..., s_j^p)$ $(1 \leq j \leq m)$, and $C = (c_1, c_2, ..., c_p)$ the centroid of $S_1, S_2, ..., S_m$. For the ideal point $O = (o^1, o^2, ..., o^p)$ in X, the distance between C and O is no greater than the average distance between m points $S_1, S_2, ..., S_m$ and O, i.e.,

$$\sqrt{\sum_{k=1}^{p} (c^k - o^k)^2} \leq \frac{1}{m} \sum_{j=1}^{m} \sqrt{\sum_{k=1}^{p} (s_j^k - o^k)^2}$$

or

$$m \sqrt{\sum_{k=1}^{p} \left(\frac{1}{m} \sum_{j=1}^{m} s_j^k - o^k\right)^2} \leq \sum_{j=1}^{m} \sqrt{\sum_{k=1}^{p} (s_j^k - o^k)^2} \tag{6.3}$$

Proof: The Minkowski Sum Inequality[2] is

$$\left(\sum_{k=1}^{p} (a_k + b_k)^q\right)^{q-1} \leq \sum_{k=1}^{p} a_k^{q-1} + \sum_{k=1}^{p} b_k^{q-1}$$

---

[2] See http://mathworld.wolfram.com/MinkowskisInequalities.html

where $q > 1$ and $a_k, b_k > 0$. For our question, we let $q = 2$ and $a_k = s_j^k - o^k$. Then we have:

$$\sum_{j=1}^{m} \sqrt{\sum_{k=1}^{p} (s_j^k - o^k)^2} = \sum_{j=1}^{m} \sqrt{\sum_{k=1}^{p} (a_k)^2}$$

$$\geq \sqrt{\sum_{k=1}^{p} (a_1 + a_2)^2} + \sum_{j=3}^{m} \sqrt{\sum_{k=1}^{p} (a_k)^2}$$

$$\geq \ldots \geq \sqrt{\sum_{k=1}^{p} (a_1 + a_2 + \ldots + a_m)^2} = \sqrt{\sum_{k=1}^{p} \left(\sum_{j=1}^{m} a_j\right)^2}$$

Now notice the left side of Inequation 6.3

$$m \sqrt{\sum_{k=1}^{p} \left(\frac{1}{m} \sum_{j=1}^{m} s_j^k - o^k\right)^2} = \sqrt{\sum_{k=1}^{p} \left(\sum_{j=1}^{m} s_k^j - m o^k\right)^2}$$

$$= \sqrt{\sum_{k=1}^{p} \left(\sum_{j=1}^{m} a_j\right)^2} \qquad \square$$

Theorem 6.3.1 tells us that the fused result by the centroid-based method is at least as effective as the average performance of all component results. Obviously, the possible maximum limit for the fused result is the ideal point. Theoretically, both maximum limit and minimum limit are achievable in certain situations. Let us illustrate this by a simple example, which comprises two component results in a two-dimensional space and is shown in the following figure.

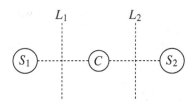

S_1 and S_2 are two component results. C is the centroid of S_1 and S_2. L_1 is such a straight line that all the points on L_1 have an equal distance to S_1 and C. L_2 is such a straight line that all the points on L_2 have an equal distance to S_2 and C. If the ideal point O is located in the area between L_1 and L_2, then the fused result F is better than both S_1 and S_2; if O is on L_1, then F is as good as S_1 and better than S_2; if O is on L_2, then F is as good as S_2 and better than S_1; if O is on the left side of L_1, then F is not as good as S_1, but better than S_2; if O is on the right side of L_2, then F is not as good as S_2, but better than S_1.

Example 6.3.1. L_1 = <$(d_1, 0.7), (d_3, 0.6), (d_2, 0.4)$>, L_2 = <$(d_2, 0.7), (d_3, 0.6), (d_1, 0.3)$>, and the fused result F = <$(d_3, 0.6), (d_2, 0.55), (d_1, 0.5)$>. Among them, d_1

and d_2 are two relevant documents, while the others are irrelevant ones. If we use ranking-based measures AP and RP to evaluate them, F is not as effective as L_1 and L_2. Now we use the Euclidean distance to evaluate them. Since documents are divided into two categories: relevant and irrelevant, we assign a relevance score of 1 to relevant documents and 0 to irrelevant documents. Therefore, we have

$$dist(L_1, O) = \sqrt{(0.7-1)^2 + (0.6-0)^2 + (0.4-1)^2} = 0.9000$$

$$dist(L_2, O) = \sqrt{(0.7-1)^2 + (0.6-0)^2 + (0.3-1)^2} = 0.9695$$

$$dist(F, O) = \sqrt{(0.6-0)^2 + (0.55-1)^2 + (0.5-1)^2} = 0.7826$$

Since $0.7826 < 0.9000$ and $0.7826 < 0.9695$, the fused result F is more effective than both component results L_1 and L_2 when the Euclidean distance is used for retrieval evaluation.   □

Example 6.3.2. Let us reconsider Example 6.1.2 using the Euclidean distance. Therefore, we have

$$dist(F, O) < dist(L_2, O)$$

$$dist(F, O) > dist(L_1, O)$$

and

$$dist(F, O) < (dist(L_1, O) + dist(L_2, O))/2$$

Therefore, F is more effective than L_2, but less effective than L_1, but more effective than the average of L_1 and L_2.   □

Theorem 6.3.2. Suppose that $S_1, S_2,..., S_m$ and O are known points in a p-dimensional space $X(q, D)$. The distance between S_i and S_j is $dist(S_i, S_j)$, the distance between O and any S_i is $dist(S_i, O)$, and the centroid of $S_1, S_2,..., S_m$ is C, then the distance between C and O can be represented as

$$dist(C, O) = \frac{1}{m} \sqrt{m \sum_{i=1}^{m} dist(S_i, O)^2 - \sum_{i=1}^{m-1} \sum_{j=i+1}^{m} dist(S_i, S_j)^2} \qquad (6.4)$$

Proof. let us assume that $O=(0,...,0)$, which can always be done by coordinate transformation. According to the definition,

$$C = (\frac{1}{m} \sum_{i=1}^{m} s_i^1, ..., \frac{1}{m} \sum_{i=1}^{m} s_i^p)$$

$$dist(C,O) = \sqrt{(\frac{1}{m}\sum_{i=1}^{m}s_i^1)^2 + \dots + (\frac{1}{m}\sum_{i=1}^{m}s_i^p)^2}$$

$$= \frac{1}{m}\sqrt{\sum_{i=1}^{m}\sum_{j=1}^{p}(s_i^j)^2 + 2\sum_{k=1}^{p}\sum_{i=1}^{m-1}\sum_{j=i+1}^{m}s_i^k * s_j^k}$$

(6.5)

Note that the distance between S_i and S_j is

$$dist(S_i,S_j) = \sqrt{\sum_{k=1}^{p}(s_i^k - s_j^k)^2}$$

or

$$dist(S_i,S_j)^2 = \sum_{k=1}^{p}s_i^{k2} + \sum_{k=1}^{p}s_j^{k2} - 2\sum_{k=1}^{p}s_i^k * s_j^k$$

Since $dist(S_i,O)^2 = \sum_{k=1}^{p}(s_i^k)^2$ and $dist(S_j,O)^2 = \sum_{k=1}^{p}(s_j^k)^2$, we get

$$2\sum_{k=1}^{p}s_i^k * s_j^k = dist(S_i,O)^2 + dist(S_j,O)^2 - dist(S_i,S_j)^2$$

Considering all possible pairs of points S_i and S_j and we get

$$2\sum_{k=1}^{p}\sum_{i=1}^{m-1}\sum_{j=i+1}^{m}s_i^k * s_j^k = (m-1)\sum_{i=1}^{m}dist(S_i,O)^2 - \sum_{i=1}^{m-1}\sum_{j=i+1}^{m}dist(S_i,S_j)^2$$

In Equation 6.5, we use the right side of the above equation to replace

$$2\sum_{k=1}^{p}\sum_{i=1}^{m-1}\sum_{j=i+1}^{m}s_i^k * s_j^k$$

Also note that $\sum_{i=1}^{m}\sum_{j=1}^{p}(s_i^j)^2 = \sum_{i=1}^{m}dist(S_i,O)^2$, we obtain

$$dist(C,O) = \frac{1}{m}\sqrt{m\sum_{i=1}^{m}dist(S_i,O)^2 - \sum_{i=1}^{m-1}\sum_{j=i+1}^{m}dist(S_i,S_j)^2} \qquad \square$$

Theorem 6.3.2 tells us that the performance of the fused result by the centroid-based method ($dist(C, O)$) can be represented by the dissimilarity among all component results and all component results' performance, or $dist(S_i, S_j)$ and $dist(S_i, O)$ for $(1 \leq i \leq m)$ and $(1 \leq j \leq m)$. To make $dist(C, O)$ as short as possible, it requires that $dist(S_i, O)$ should be as short as possible and $dist(S_i, S_j)$ should be as long as possible. That is, all component results should be as effective as possible, and all pairs of component results should be as different as possible. On the other hand, we can work out the effectiveness of the fused result without fusing them at all, if the effectiveness of all component results and the dissimilarity between them are known.

Example 6.3.3. Let $m = 2$, $dist(S_1, O) = 12$, $dist(S_2, O) = 10$, $dist(S_1, S_2) = 8$.

$$dist(C,O) = \frac{1}{2}\sqrt{2 \times dist(S_1,O)^2 + 2 \times dist(S_2,O)^2 - dist(S_1,S_2)^2}$$
$$= \frac{1}{2}\sqrt{2 \times (12^2 + 10^2) - 8^2} = 10.30 \qquad \square$$

Theorem 6.3.3. In X, there are two groups of points: $\mathscr{S}_1 = \{ S_{11}, S_{21},..., S_{m1}\}$ and $\mathscr{S}_2 = \{ S_{12}, S_{22},..., S_{m2}\}$. C_1 is the centroid of $S_{11}, S_{21},..., S_{m1}$ and C_2 is the centroid of $S_{12}, S_{22},..., S_{m2}$. O is the ideal point. Suppose $\overline{dist_1} = \frac{1}{m}\sum_{j=1}^{m} dist(S_{j1},O)$ and $\overline{dist_2} = \frac{1}{m}\sum_{j=1}^{m} dist(S_{j2},O)$, $v_1 = \frac{1}{m-1}\sum_{j=1}^{m}(dist(S_{j1},O) - \overline{dist_1})^2$ and $v_2 = \frac{1}{m-1}\sum_{j=1}^{m}(dist(S_{j2},O) - \overline{dist_2})^2$. Here v_1 is the variance of $dist(S_{j1},O)$ ($j = 1, 2,...,$ m) and v_2 is the variance of $dist(S_{j2},O)$ ($j = 1, 2,..., m$). And the following three conditions hold:

$$(a) \sum_{k=1}^{m-1} \sum_{l=k+1}^{m} dist(S_{k1},S_{l1}) = \sum_{k=1}^{m-1} \sum_{l=k+1}^{m} dist(S_{k2},S_{l2})$$
$$(b) \overline{dist_1} = \overline{dist_2}$$
$$(c) v_1 < v_2$$

then we have $dist(C_1, O) < dist(C_2, O)$.

Proof: according to v_1' and v_2' definitions,

$$v_1 = \frac{1}{m-1}\sum_{j=1}^{m}(dist(S_{j1},O) - \overline{dist_1})^2 = \frac{1}{m-1}\sum_{j=1}^{m} dist(S_{j1},O)^2 - \frac{m-2}{m-1}\overline{dist_1}^2$$
$$(6.6)$$

$$v_2 = \frac{1}{m-1}\sum_{j=1}^{m}(dist(S_{j2},O) - \overline{dist_2})^2 = \frac{1}{m-1}\sum_{j=1}^{m} dist(S_{j2},O)^2 - \frac{m-2}{m-1}\overline{dist_2}^2$$
$$(6.7)$$

Comparing Equation 6.6 with 6.7 and applying conditions (b) and (c), we have

$$\sum_{j=1}^{m} dist(S_{j1},O)^2 < \sum_{j=1}^{m} dist(S_{j2},O)^2 \qquad (6.8)$$

Considering Inequation 6.8 and Condition (a) in $dist(C_1, O)$'s and $dist(C_2, O)$'s definitions, we conclude that $dist(C_1, O) < dist(C_2, O)$. $\qquad \square$

Theorem 6.3.3 tells us that the variance of effectiveness of all component results affect the centroid-based method when all other conditions are the same. The smaller the variance is, the better result we can obtain. On the other hand, unlike effectiveness, only the total dissimilarity, not the variance of dissimilarity among all component results matters. This can be ascertained from Equation 6.4.

Theorem 6.3.4. In X, S_1, S_2,..., S_m and the ideal point O are known points. The centroid of S_1, S_2,..., S_m is C. There are m different sets of points, each of which includes m-1 points from S_1, S_2,..., S_m. The centroid of S_2, S_3,..., S_m is C_1; the centroid of S_1, S_3,..., S_m is C_2; the centroid of S_1, S_2,..., S_{m-1} is C_m. We have

$$dist(C,O) < \frac{1}{m} \sum_{i=1}^{m} dist(C_i,O) \qquad (6.9)$$

Proof: according to Theorem 6.3.2, we have

$$dist(C,O)^2 = \frac{1}{m^2}[m \sum_{j=1}^{m} dist(s_j,O)^2 - \sum_{k=1}^{m-1} \sum_{l=k+1}^{m} dist(S_k,S_l)^2]$$

Now let us look at the right side of Inequation 6.9. Also according to Theorem 6.3.2, we obtain

$$\frac{1}{m} \sum_{i=1}^{m} dist(C_i,O)^2$$

$$= \frac{1}{m(m-1)^2}[(m-1) \sum_{i=1}^{m} \sum_{j=1 \wedge j \neq i}^{m} dist(S_j,O)^2 - \sum_{i=1}^{m} \sum_{j=1 \wedge j \neq i}^{m-1} \sum_{k=j+1 \wedge k \neq i}^{m} dis(S_j.S_k)^2]$$

$$= \frac{(m-1)^2 \sum_{j=1}^{m} dist(S_j,O)^2 - (m-2) \sum_{j=1}^{m-1} \sum_{k=j+1}^{m} dist(S_j,S_k)^2}{m(m-1)^2}$$

$$= \frac{1}{m^2}[\sum_{j=1}^{m} dist(S_j,O)^2 - \frac{m(m-2)^{2(m-1)}}{m-1} \sum_{j=1}^{m} \sum_{k=j+1}^{m} dist(S_j,S_k)^2]$$

$$> \frac{1}{m^2}[\sum_{j=1}^{m} dist(S_j,O)^2 - \sum_{j=1}^{m-1} \sum_{k=j+1}^{m} dist(S_j,S_k)^2]$$

$$= dist(C,O)^2 \qquad \square$$

Theorem 6.3.4 can be proved in more general situations in which a subset includes m-2, or m-3,..., or 2 points. This is to say, the distance between the ideal point and the centroid of a group of m points is shorter than the average distance between the ideal point and the centroids of $\frac{m!}{p!(m-p)!}$ groups, each of which comprises $(m-p)$ points . This demonstrates that in general better performance is achievable if we fuse more component results.

If no information is available about the performances of all component results, then the best assumption is that all of them are equally effective and the centroid-based data fusion method is the best data fusion method. We shall discuss this in Section 6.4.

6.4   The Linear Combination Method

The linear combination method is a more flexible method than the centroid-based method. For a group of results (points) $S_j=(s_{j1},, s_{jp})$ $(1 \leq j \leq m)$, the linear combination method uses the following equation to calculate scores:

$$f^k = \sum_{j=1}^{m} w_j s_j^k \qquad (6.10)$$

where w_j is the weight assigned to each information retrieval system ir_j and $F = (f^1, f^2,..., f^p)$ is the fused result.

If the performances of all component results are known, then the linear combination method is very likely more effective than the centroid-based method. However, if no such information is available, then we may have two options. One is to use the centroid-based method (equal weights for all component results), the other is to use the linear combination with different weights that are randomly selected for each of the component results. Sometimes the second option is better than the first option, sometimes vice versa. However, on average the centroid-based option is better. Therefore, the centroid-based method can be regarded as the best when no such information is available. Theorem 6.4.1 proves this.

Theorem 6.4.1. In a p-dimensional space X, there are m points $S_j=(s_j^1,..., s_j^p)$ $(1 \leq j \leq m)$ and the ideal point $O = (o^1,.., o^p)$. $C = (c^1,..., c^p)$ is the centroid of $S_1, S_2,..., S_m$ and the distance between C and O is $dist(C, O)$. $F = w_1 S_1 + w_2 S_2 +...+ w_m S_m$ is the linear combination of $S_1, S_2,..., S_m$, here $w_1 + w_2 +...+ w_m = 1$. Then for all possible $w_1, w_2,..., w_m$, the average distance between F and O is no less than $dist(C, O)$.

Proof: the number of different linear combinations of $S_1, S_2,..., S_m$ is indefinite. We do not try to list and evaluate all of them. Instead, we divide them into groups, each of which includes m combinations. For any valid combination $w_1 S_1 + w_2 S_2 +...+ w_m S_m$, we put other $m-1$ combinations $w_2 S_1 + w_3 S_2 +...+ w_1 S_m$, $w_3 S_1 + w_4 S_2 +...+ w_2 S_m,..., w_m S_1 + w_1 S_2 +...+ w_{m-1} S_m$, into the same group. We can calculate that the centroid of these m combinations is C (note that $w_1 + w_2 +...+ w_m = 1$), same as the centroid of $S_1,..., S_2,..., S_m$. Therefore, according to Theorem 6.3.1, we can derive that the average distance between these m points $w_1 S_1 + w_2 S_2 +...+ w_m S_m$, $w_2 S_1 + w_3 S_2 +...+ w_1 S_m,..., w_m S_1 + w_1 S_2 +...+ w_{m-1} S_m$ and O is no less than $dist(C, O)$.

Finally, we are certain that any valid combination can be in one and only one of such groups. Thus the proof is complete.   □

Theorem 6.4.2. In a p-dimensional space X, there are m points $\mathscr{S}_1 = (S_1, S_2,..., S_m)$ $(1 \leq j \leq m)$ and the ideal point is O. $W_1=(w_{11}, w_{12},..., w_{1m})$ is a group of weights and $F_1 = w_{11} S_1 + w_{12} S_2 +...+ w_{1m} S_m$. W_1 is the optimum weight for $\mathscr{S}_1$ since it makes F_1 have the shortest distance to O ($dist(F_1, O)$) among all such weights. Now we add one more point S_{m+1} to $\mathcal{S}_1$ to form $\mathcal{S}_2$, or $\mathcal{S}_2 = (S_1,, S_m, S_{m+1})$, and W_2 $= (w_{21}, w_{22},\dots, w_{2m}, w_{2(m+1)})$ is the optimum weights for $\mathcal{S}_2$ to make $F_2 = w_{21}S_1$ $+ w_{22}S_2 +\dots+ w_{2m}S_m + w_{2(m+1)}S_{m+1}$ have the shortest distance to O ($dist(F_2, O)$) among all such weights. Then we have $dist(F_1, O) \geq dist(F_2, O)$.

Proof: if we let $w_{2(m+1)} = 0$, and $w_{21} = w_{11}, w_{22} = w_{12},\dots, w_{2m} = w_{1m}$. Then $F_2 = F_1$ and $dist(F_2, O) = dist(F_1, O)$. Therefore, $dist(F_2, O)$ is at least as short as $dist(F_1, O)$. This is the case if S_{m+1} can be linearly represented by the other m points.

To make the proof complete, we only need to point out one example in which $dist(F_2, O) < dist(F_1, O)$: let us consider $S_1 = (2, 0)$, $S_2 = (0, 2)$, $O = (0, 0)$, the shortest possible distance between O and a linear combination of S_1 and S_2 is $\sqrt{2}$; if we add a third point $S_3 = (0, 1)$, then the shortest possible distance between O and a linear combination of S_1, S_2, and S_3 is 1. □

Note that we cannot obtain similar conclusions for the centroid-based method. If a very poor result is added, then the effectiveness of the fused result will be affected significantly. Therefore, very poor results should be avoided when using the centroid-based method. On the other hand, this theorem demonstrates the flexibility of the linear combination method. No matter how poor a result can be, it will do no harm at all to fusion effectiveness if proper weights can be assigned to all component results.

Theorem 6.4.3. In a p-dimensional space X, there are m points $\mathcal{S} = (S_1, S_2,\dots, S_m)$ for ($1 \leq i \leq m$) and the ideal point is O. $W = (w_1, w_2,\dots, w_m)$ is a group of optimum weights that make the distance between $F = \sum_{i=1}^{m} w_i S_i$ and O ($dist(F, O)$) the shortest. Then $dist(F, O) \leq \min(dist(S_1, O), dist(S_2, O),\dots, dist(S_m, O))$ must hold. Here $\min(a_1, a_2,\dots, a_m)$ is the function that takes the smallest in ($a_1, a_2,\dots, a_m$) as its value.

Proof: suppose $dist(S_i, O)$ is the smallest among all $dist(S_j, O)$ for ($1 \leq j \leq m$). Then we can set a weight of 1 for S_i and 0 for all other points. With such a group of weights, $dist(F, O) = dist(S_i, O)$. In most situations, we are able to find many linear combinations of $S_1, S_2,\dots, S_m$, whose distance to O is shorter than $dist(S_i, O)$. For example, if $S_1 = (1, 0)$, $S_2 = (0, 1)$, $S_3 = (-2, 0)$, $O = (0, 0)$, then $dist(S_1, O) = 1$, $dist(S_2, O) = 1$, $dist(S_3, O) = 2$. If we let $F = S_1 + 0.5 * S_3$, then $dist(F, O) = 0$. Thus $dist(F, O) < dist(S_i, O)$ for ($1 \leq i \leq 3$). □

This theorem tell us that if only considering one query and the optimum weights are used, then the fused result using the linear combination method is at least as effective as the most effective component result. This is really a good property. However, if multiple queries are considered together, the same conclusion does not held theoretically. As a matter of fact, for any information retrieval system, the result generated can be seen as the reflection of the characteristics of the information retrieval system involved. If we always use a certain group of information retrieval systems to generate results for fusion, then these results generated over different queries should have quite stable relations (or similarities) with each other. Therefore, in most cases, we can expect that the fused result using the linear combination method with the optimum weights is more effective than the best component result even we consider multiple queries together.

A practical problem is how to decide the optimum weights for a group of queries. Suppose we have a collection of p documents $D = (d_1, d_2,..., d_p)$, a group of information retrieval systems $IR = (ir_1, ir_2,..., ir_m)$, and a group of queries $Q = (q_1, q_2,..., q_n)$. For each query q_i, every retrieval system ir_j returns a set of scores $S_{ij} = (s_{ij}^1, s_{ij}^2,..., s_{ij}^p)$[3], which are relevant scores of documents $D = (d_1, d_2,..., d_p)$. $O_i = (o_i^1, o_i^2,..., o_i^p)$ is the real relevance of all the documents for query q_i $(1 \leq i \leq n)$. The problem is to find a group of weights $(w_1, w_2,..., w_m)$ to minimize the following mathematical expression:

$$\sum_{i=1}^{n} \sqrt{\sum_{k=1}^{p} (\sum_{j=1}^{m} w_j S_{ij}^k - O_{ij}^k)^2} \qquad (6.11)$$

Since it is very difficult to work out the solution for the above mathematical expression, we can use the following expression instead based on the principle of least squares:

$$\sum_{i=1}^{n} \sum_{k=1}^{p} (\sum_{j=1}^{m} w_j S_{ij}^k - O_{ij}^k)^2 \qquad (6.12)$$

Let

$$f(w_1, w_2, ..., w_m) = \sum_{i=1}^{n} \sum_{k=1}^{p} (\sum_{j=1}^{m} w_j S_{ij}^k - O_{ij}^k)^2$$

then, we have

$$\frac{\partial f}{\partial w_q} = \sum_{i=1}^{n} \sum_{k=1}^{p} 2(\sum_{j=1}^{m} w_j S_{ij}^k - O_i^k) S_{iq}^k = 0$$

or

$$\sum_{j=1}^{m} w_j (\sum_{i=1}^{n} \sum_{k=1}^{p} S_{iq}^k S_{ij}^k) = \sum_{i=1}^{n} \sum_{k=1}^{p} O_i^k S_{iq}^k$$

for $q = 1, 2,..., m$. Let

$$a_{qi} = \sum_{i=1}^{n} \sum_{k=1}^{p} S_{iq}^k S_{ij}^k$$

and

$$d_q = \sum_{i=1}^{n} \sum_{k=1}^{p} O_i^k S_{iq}^k$$

---

[3] Here we assume that those score vectors are independent. This is almost always the case for practical applications in which those results are from different retrieval systems. However, theoretically, if there are any dependent score vectors, then we need to remove them in the first place.

Thus we obtain the following m equations in m variables:

$$\begin{pmatrix} a_{11}w_1 + a_{12}w_2 + \cdots + a_{1m}w_m = d_1 \\ a_{21}w_1 + a_{22}w_2 + \cdots + a_{2m}w_m = d_2 \\ \ldots\ldots\ldots\ldots\ldots\ldots\ldots\ldots\ldots\ldots\ldots \\ a_{m1}w_1 + a_{m2}w_2 + \cdots + a_{mm}w_m = d_m \end{pmatrix}$$

Thus, the optimum weights[4] can be calculated by finding the solution of these m linear equations.

In most cases, we can expect that the linear combination method, with the optimum weights, is more effective than the centroid-based method. If the same weights are used for a group of information retrieval systems across a group of queries, then the effectiveness of the linear combination method depends on the consistency of effectiveness of all component retrieval systems and the consistency of similarity among all component results. If both system performance and similarity among results are very consistent over all the queries, then the linear combination method is very effective; otherwise, if either or both or them varies considerably from one query to another, then the linear combination method becomes less effective. The worst case is that all systems have to be assigned the same weight and the linear combination method equals to the centroid-based method.

6.5 Relation between the Euclidean Distance and Ranking-Based Measures

In Sections 6.2-6.4 we have discussed the geometric framework, in which the Euclidean distance was used. A number of properties were presented. However, in practice, some ranking-based metrics are very often used. In this section we investigate the relationship between the Euclidean distance and some commonly used ranking-based measures, so as to determine the degree to that all those properties of fusion will be held when ranking-based measures are used.

We can prove that all properly defined ranking-based measures, such as average precision, recall level precision, precision at curtain cut-off document level, normalized discount cumulative gain (NDCG), and others, are reasonable (see Section 2.1), though some of them are more sensitive than the others. For example, suppose we use binary relevance judgment to evaluate two results $L_1 = <r, i, i, r, i>$ and $L_2 = <i, r, i, r, i>$ and there are only 2 relevant documents in the whole collection. Here 'r' denotes a relevant document and 'i' denotes an irrelevant document. L_1 and L_2 are only different at rank 1 and rank 2. If we use AP, then $AP(L_1) = (1/1 + 2/4)/2 = 0.75$ and $AP(L_2) = (1/2 + 2/4)/2 = 0.5$. L_1 is assigned a higher value than L_2. If we use P@5 (precision at 5 document level), then $P@5(L_1) = P@5(L_2) = 2/5 = 0.4$.

---

[4] The weights calculated are optimum for Equation 6.12. They are not necessarily optimal for Equation 6.11. However, we still refer to such weights as optimal, in the sense of least squares. This is widely accepted in statistics [34], operations research, and other fields.

The difference between L_1 and L_2 does affect AP but not P@5. This example illustrates that P@5 is not as sensitive as AP, though both of them are reasonably defined measures.

Theorem 6.5.1. Using any ranking-based measure that are reasonably defined, the best way of ranking a group of documents is to rank them according to their relevance scores from the highest to the lowest.

This theorem can be deduced directly from the four rules of reasonably defined measures (see Section 2.1)   □

For a group of documents D, we can generate one or more ranked list of them by applying Theorem 6.5.1. For any such ranked list of documents $<d_1, d_2,..., d_n>$, $o(d_1) \geq o(d_2) \geq...\geq o(d_n)$. Here $o(d_i)$ is the ideal relevance score of d_i. If multiple documents have equal scores, then we can obtain multiple equally effective rankings of documents. Such ranking(s) are referred to as the optimum ranking(s) of D.

The Euclidean distance is a measure that considers all the documents in the result. $dist(S,O) = \sqrt{\sum_{i=1}^{p} (s^i - o^i)^2}$ for $S = (s^1, s^2,..., s^p)$ and $O = (o^1, o^2,..., o^p)$. For every element s^i in S, we can define its distance from the corresponding element (document) o^i: $dist(s^i, o^i) = \sqrt{(s^i - o^i)^2} = |s^i - o^i|$. Here $dist(s^i, o^i)$ is referred to as the distance of the i-th element (document) between S and O.

Theorem 6.5.2. $O = (o^1, o^2,..., o^p)$ is the ideal point in a p ($p \geq 2$) dimensional space X. Here each o^i is the accurate relevance value of document d_i ($1 \leq i \leq p$). $\mathcal{G}_1$ is a collection of points in X, which comprise all those points that have an equal distance of a_1 to O. $\mathcal{G}_2$ is another collection of points in X, which comprises all those points that have an equal distance of a_2 to O. $a_1 < a_2$. Then for any dimension i, the average distance of the i-th element between the point in $\mathcal{G}_1$ and O is shorter than the average distance of the i-th element between the point in $\mathcal{G}_2$ and O.

Proof: all the points in $\mathcal{G}_1$ form a super-spherical surface with center O and radius a_1; while all the points in $\mathcal{G}_2$ form a super-spherical surface with center O and radius a_2. For any point G_{11} in $\mathcal{G}_1$, we can find a corresponding point G_{21} in $\mathcal{G}_2$: we connect O and G_{11} by a straight line and extend it to point G_{21} in $\mathcal{G}_2$, and make O and G_{11} and G_{21} on a straight line. On the other hand, for any point G_{21} in $\mathcal{G}_2$, we can also find a corresponding G_{11} in $\mathcal{G}_1$ that makes O and G_{11} and G_{21} on a straight line. In both cases, for any dimension i ($1 \leq i \leq p$), the distance of the i-th element between G_{11} and O is shorter than the distance of the i-th element between G_{21} and O. Since such a mapping is one to one and can be carried out for any points in $\mathcal{G}_1$ and $\mathcal{G}_2$, the theorem is proved.   □

Theorem 6.5.2 tells us that considering all the points in a space, if a group of equally effective results $\mathcal{G}_1$ is more effective than another group of equally effective results $\mathcal{G}_2$, then on average the relevance score of every document in $\mathcal{G}_1$ is more accurately estimated than that in $\mathcal{G}_2$.

For any two different rankings of the same group of documents, we can always transform one to the other by one or more neighbouring swaps. For example, suppose $L_1 = <d_1, d_2, d_3, d_4>$ and $L_2 = <d_2, d_3, d_1, d_4>$ are two ranked lists of the same group of documents. We can transform L_2 to L_1 by swapping d_1 with d_3, and then d_1 with d_2.

Definition 6.5.1. In a ranked list of documents $L = <d_1, d_2,.., d_n>$, d_i and d_{i+1} are two neighbouring documents, where $(1 \leq i < i+1 \leq n)$. The ideal relevance score of d_i is $o(d_i)$ and the ideal relevance score of d_{i+1} is $o(d_{i+1})$. We say a swap of d_i and d_j is beneficial if $o(d_i) < o(d_{i+1})$. $\square$

Theorem 6.5.3. $O = (o^1, o^2,..., o^p)$ is the ideal point in a p $(p \geq 2)$ dimensional space X. $opti(O)$ is the optimum ranking related to O. $\mathcal{G}_1$ is a collection of points in X, which comprise all those points that have an equal distance of a_1 to O. $\mathcal{G}_2$ is another collection of points in X, which comprises all those points that have an equal distance of a_2 to O. $a_1 < a_2$. Then on average, transforming $opti(G_{2i})$ to $opti(O)$ for any $G_{2i} \in \mathcal{G}_2$ requires at least as many beneficial neighbouring swaps as transforming $opti(G_{1j})$ to $opti(O)$ for any $G_{1j} \in \mathcal{G}_1$. $\square$

This theorem can be proved in a very similar way as Theorem 6.5.2. Theorem 6.5.3 shows that a positive correlation exists between the number of beneficial neighbouring swaps and $dist(O, S)$.

The performance of a ranked list of documents can be evaluated by how many beneficial neighbouring swaps it takes to transform that ranking to the optimum ranking or one of the optimum rankings. Obviously, the number of beneficial neighbouring swaps as a measure of effectiveness is reasonable. And it is one of the most sensitive measures, since any beneficial neighbouring swap will affect its value.

If D is a collection of p documents, then we can expect that for any ranking of D, $L = <d_1, d_2,..., d_p>$, the number of beneficial neighbouring swaps it needs to transform to the optimum ranking is strongly correlated to the value of any reasonable ranking-based measures if a large number of cases are considered. This is supported by the following observations:

- If the ranking considered is optimum, then no beneficial neighbouring swap is needed and all reasonable measures obtain their best values. If the ranking considered is the worst, then a maximum number of beneficial neighbouring swaps are needed and all reasonably defined measures obtain their worst values. In these two extreme situations, all reasonably defined measures are consistent.
- For any other ranking L of D, we can transform the worst ranking to L by a number of beneficial neighbouring swaps, and we can also transform L to the optimum ranking by a number of beneficial neighbouring swaps. Assume that W L_1 L_2 ... L_q L L_{q+1} L_{q+2} .. L_r O is the series of rankings we go through when swapping neighbours from the worst to the optimum ranking. Here W denotes the worst ranking, O denotes the optimum ranking. In this series of rankings, the closer to O the ranking is, the more effective the ranking is. If L_a and L_b are two different rankings in such a series and L_a is closer to O than L_b, then L_a will be evaluated more effective than or at least as effective as L_b by any reasonable ranking-based measure.

- For two different series of rankings of the same group of documents: $W\ L_{11}\ L_{12}...$ $L_{1q}\ O$ and $W\ L_{21}\ L_{22}\ ...\ L_{2r}\ O$, if L_{1i} is more effective than L_{2j} by any reasonably defined measure, then $L_{1(i+1)}, L_{1(i+2)},..., L_{1q}$ are more effective than L_{2j} by the same measure. If L_{1i} is less effective than L_{2j} by any reasonably defined measure, then $L_{11}, L_{12},..., L_{1(i-1)}$ are less effective than L_{2j} by the same measure. Therefore, we can expect that any reasonable ranking-based measure is negatively correlated to the number of beneficial neighbouring swaps. We can also expect that any two reasonable ranking-based measures are strongly positively correlated.

Up to this point, we have established the relationship between the Euclidean distance and all reasonable ranking-based measures using the number of beneficial neighbouring swaps as an intermediate. We have also demonstrated that if a large number of points are considered, a negative correlation between the Euclidean distance and any reasonable ranking-based measures should exist. Next let us discuss the relationship between neighbouring documents swap and precision at first k ($k \geq 1$) document level (P@k). Precision at first k document level is one of commonly used measures in retrieval evaluation. Suppose that binary relevance judgment is used and there are m relevant documents in all n documents. Furthermore, we assume that at each ranking position, the probability that a document is relevant is linearly related to its rank:

$$Pr(t) = \frac{m(n-t+1)}{0.5n(n+1)}\ for\ (1 \leq t \leq n)$$

then

$$\sum_{t=1}^{n} Pr(t) = \sum_{t=1}^{n} \frac{m(n-t+1)}{0.5n(n+1)} = \frac{m(n+n-1+...+1)}{0.5n(n+1)} = m$$

The probability that a document at rank t is irrelevant is $1\text{-}Pr(t)$. For a total number of n ranked documents, there are n-1 possible swaps of neighbouring document pairs. We assume that for each pair of them the sway happens in an equal probability $1/(n\text{-}1)$.

If a swap happens between documents d_k and d_{k+1}, d_k is an irrelevant document and d_{k+1} is a relevant document, then P@k will increase by $1/k$. Therefore, the probability that any neighbouring documents swap may increase the value of P@k by $1/k$ is $\frac{1}{n-1}(1 - Pr(k))Pr(k+1)$.

Other metrics such as average precision, recall-level precision, the reciprocal rank, can also be dealt with in a similar way.

6.6   Conclusive Remarks

In this chapter we have formally discussed score-based data fusion methods under a geometric framework, in which each result is represented by a point in the result space and both performance and similarity between results can be measured by the same metric, the Euclidean distance. The benefit of it is to make data fusion a deterministic problem. This is very good for us to understand the properties of the data fusion methods.

However, usually ranking-based metrics are used for retrieval evaluation. The connection between score-based metrics and ranking-based metrics is a key issue, which decides how and to what extent we can apply all the useful theorems and properties obtained from the former to the latter. Although some discussion about this has been carried out in Section 6.5, further investigation is desirable. One hypothesis is: we can expect that all the properties of the data fusion methods obtained in the geometric framework still make sense for ranking-based metrics, if we consider a large number of data fusion instances collectively and all the instances are taken randomly from the whole result space. The investigation can be carried out in either or both directions: theoretically or empirically.

Chapter 7
Ranking-Based Fusion

In information retrieval, retrieval results are usually presented as a ranked list of documents for a given information need. Thus ranking-based fusion methods are applicable even no scoring information is provided for all the documents involved. In this chpater, we are going to investigate ranking-based fusion methods especially the Borda count, the Condorcet voting and the weighted Condorect voting.

7.1 Borda Count and Condorcet Voting

In political science, the Borda count and the Condorcet voting are two well-known voting methods. It has been argued for long time which of these two methods is better. There is no consensus because different people have different opinions of what a good voting system means to them and there is no universally accepted criterion.

The Borda count and the Condorcet voting can be used for data fusion in information retrieval without any problem. For all data fusion methods in information retrieval, we are able to evaluate them to see which one is good and which one is not by theoretical analysis or/and experiments. This will make some conclusive remarks for a comparison of them much less controversial.

After assigning scores to documents at different ranks (for Borda count) or normalizing raw scores (for CombSum), Both Borda count and CombSum use the same equation (see Equation 1.1 in Section 1.2) to calculate scores for all the documents involved. Therefore, they bear some similarity. Most previous discussions of CombSum are also applicable to the Borda count as well. For example, the various methods discussed in Chapter 5 can be used to extend both CombSum and Borda count to linear combination in more or less the same way.

From the discussion before, we can see that the only difference between the Borda count and CombSum is how to generate the scores for those documents occurred in component results. From the discussion in Chapter 3, we conclude that the method used by the Borda count is not as good as other methods such as the logistic model (see Section 3.3.4) or the cubic model (see Section 3.3.5).

S. Wu: Data Fusion in Information Retrieval, ALO 13, pp. 135–147.
springerlink.com   © Springer-Verlag Berlin Heidelberg 2012

Therefore, the Borda count does not seem to be a very promising data fusion method in information retrieval.

Condorcet voting is quite different from some other fusion methods such as CombSum and CombMNZ in the process of deciding the rankings of all the documents involved. But just like the others, it can still be beneficial from assigning different weights to component systems so as to archive better fusion results. Especially when the retrieval systems involved are very different in performance or the similarity between some of the results are very uneven, it is very desirable to use weighted Condorcet voting to replace Condorcet voting for better fusion performance.

The problem of uneven similarity between some component results may cause more damage to Condorcet than to the Borda count. Let us see two examples.

Example 7.1.1. Suppose that there are $2n + 1$ component systems, and the ranked lists from $n + 1$ of them are exactly the same (denoted as L), then the fused result by Condorcet voting is the same as L.

This can be proven by considering any pair of documents (d_i, d_j) in the lists. Since $n + 1$ ranked lists are just the same, let us assume that d_i is ranked ahead of d_j, or $< d_i, d_j >$, by all of them. The final decision is $< d_i, d_j >$ no matter how the other n lists rank d_i and d_j. This example is very good for illustrating one important property of the Condorcet voting: dictatorship of the majority.   □

For the above exmaple, if we use CombSum or Borda count to fuse them, then it is very unlikely that we can obtain a fusion result like this. In an extreme situation, the relative ranking of two documents may only be decided by one retrieval system.

Example 7.1.2. Suppose that there are n component systems, each of which provides a ranked list of m documents $(m > n)$. In $n - 1$ ranked lists, d_1 is ranked one place higher than d_2; only in one ranked list, d_2 is ranked in the first place, while d_1 is ranked in the last place. If we use the Borda count to fuse them, then d_2 will be ranked ahead of d_1.

The conclusion can be proven by considering the score difference between the two documents d_1 and d_2. In each of those $n - 1$ results, d_1 only obtain one more score than d_2; while in the last result, d_2 obtain $m - 1$ more scores than d_1. Therefore, d_2 is ranked ahead of d_1 because d_2 obtains more scores than d_1 ($m > n$ and $m - 1 > n - 1$).   □

Example 7.1.2 shows that in the Borda count, any individual voter (information retrieval system) may have the decisive power to the final ranking of any pair of documents. This is not possible for Condorcet voting. In Condorcet, each retrieval result has exactly the same power as others. Considering two extreme situations: one is that d_j is just next to d_i; the other is d_i is at the top while d_j is at the bottom. In the Condorcet voting, we do not distinguish such a difference and treat them equally. We may explain this as the ambiguity of the Condorcet voting method or the sensitivity of the Borda count to every component result. However, this does not necessarily mean that Condorcet is inferior to the Borda count in performance or vice versa. Detailed discussion about the properties of the Borda count and the Condorcet voting can be found in [76].

From the avove two examples, it seems that a lot of interesting properties about them are unknown to us. Definitely these two examples can not tell us the whole story. We hypothesize that: compared with some other methods such as CombSum and the Borda count, unfavourable conditions such as performance variation of component results and/or uneven dissimilarity between component results may cause Condorcet voting more performance deterioration. We shall discuss this issue in Section 7.5.

7.2 Weights Training for Weighted Condorcet Voting

In this section, we are going to discuss how to apply linear discrimination to train weights for weighted Condorcet voting. Linear discrimination is a technique that linearly separate instances into two or more categories. This approach estimates the parameters of the linear discriminant directly from a given labelled sample through a search for the parameter values that minimize an error function. This approach is efficient and both space and time complexities are $O(m)$, where m is the number of samples [1].

The key issue in Condorcet voting is the pairwise competition. If both documents involved are relevant (or irrelevant) at the same time, then how to rank them is not important. Since no matter which document wins the competition, the performance of the result will not be affected. What matters is: if a relevant document and an irrelevant document are in a pairwise competition, then we wish that the relevant document is able to win. Even by assigning different weights to different component results, it is usually not possible to let all the relevant documents be always winners. Anyway, we should try our best to let as many relevant documents win the competition as possible. This is the rationale behind the weights assignment for weighted Condorcet voting.

Suppose there are n information retrieval systems ir_1, ir_2, ..., ir_n. For a query q, each of them returns a ranked list of documents $L_j (1 \leq j \leq n)$ and D is the set of all the documents involved. For simplicity, we assume that all L_js comprise the same group of documents. This is possible, for example, if all the documents in the collection are retrieved. We shall discuss a variant of it later by removing this restriction. We may divide D into two sub-collections: relevant documents D_r and irrelevant documents D_i. There are $|D_r|$ documents in D_r and $|D_i|$ documents in D_i. If we choose one from each collection, then we have a total number of $2|D_r||D_i|$ ranked pairs. Note that $< d_a, d_b >$ and $< d_b, d_a >$ are different pairs. For all $2|D_r||D_i|$ pairs, we divide them into two classes: Class X and Class Y. Class X comprises all those pairs that a relevant document is ranked ahead of an irrelevant document, represented by '+1'; and Class Y comprises all those pairs that an irrelevant document is ranked ahead of a relevant document, represented by '-1'. For each ranked pair, we check every component result to see if the pair is supported or not. If the ranked pair $< d_a, d_b >$ is supported by ir_i, which means that d_a is also ranked ahead of d_b in L_i, then we use '+1' to represent it; If the ranked pair $< d_a, d_b >$ is not supported by ir_i, which means that d_b is ranked ahead of d_a in L_i, then we use '-1' to represent it. For all the ranked pairs, we repeat this process over all component results. Thus for each ranked pair (instance), it has n features, each of which is obtained from a component system.

Example 7.2.1. $D_r = \{d_1, d_3, d_5\}$, $D_i = \{d_2, d_4\}$, $L_1 = < d_1, d_3, d_2, d_4, d_5 >$, $L_2 = < d_2, d_1, d_3, d_5, d_4 >$, $L_3 = < d_5, d_4, d_3, d_1, d_2 >$, the following table

Number	Pair	f_1	f_2	f_3	Category
1	$< d_1, d_2 >$	+1	-1	+1	+1
2	$< d_2, d_1 >$	-1	+1	-1	-1
3	$< d_1, d_4 >$	+1	+1	-1	+1
4	$< d_4, d_1 >$	-1	-1	+1	-1
5	$< d_3, d_2 >$	+1	-1	+1	+1
6	$< d_2, d_3 >$	-1	+1	-1	-1
7	$< d_3, d_4 >$	+1	+1	-1	+1
8	$< d_4, d_3 >$	-1	-1	+1	-1
9	$< d_5, d_2 >$	-1	-1	+1	+1
10	$< d_2, d_5 >$	+1	+1	-1	-1
11	$< d_5, d_4 >$	-1	+1	+1	+1
12	$< d_4, d_5 >$	+1	-1	-1	-1

can be used to represent all the instances with their features. Let us look at 2 pairs (instances). Instance number 1 is $< d_1, d_2 >$. It is supported by ir_1 and ir_3, but not ir_2, therefore, $f_1 = +1$, $f_3 = +1$, and $f_2 = -1$. Because d_1 is a relevant document and d_2 is an irrelevant document, the instance belongs to class X and we put a '+1' for its category. Instance number 10 is $< d_2, d_5 >$. It is supported by ir_1 and ir_2, but not ir_3. Therefore, $f_1 = +1$, $f_2 = +1$, and $f_3 = -1$. This instance belongs to class Y because d_2 is an irrelevant document and d_5 is a relevant document. We put a '-1' for its category.

Now we want to distinguish the instances of the two classes by a linear combination of n features. Let

$$g(f_1, f_2, ..., f_n) = \sum_{i=1}^{n} w_i f_i + w_0 \tag{7.1}$$

If $g(f_1, f_2, ..., f_n) > 0$, the instance belongs to Class X; if $g(f_1, f_2, ..., f_n) \leq 0$, the instance belongs to Class Y. The linear discriminant analysis can be used to provide an optimised solution[1].

In this example, by using linear discriminant analysis (LDA) we obtain the weights w_1, w_2, and w_3 for f_1, f_2, and f_3 are 1.265, 1.342, and 1.897, respectively. Note that in the above table, each feature f_i (column) is obtained from a given information retrieval system ir_i, thus the weight obtained for f_i is for ir_i as well. □

---

[1] After projection from n dimensions to 1 dimension, for the two classes to be well separated, we would like the means to be as far apart as possible and the examples of classes be scattered in as small a region as possible [1].

7.3 Experimental Settings and Methodologies

In this section we present experimental settings and methodologies, and the experimental results will be presented in the next two sections. Three groups of runs submitted to TREC (8, 2003, and 2008) are used. Their information is summarized in Table 7.1. See Appendix C for detailed information about the runs involved.

Table 7.1 Summary of all three groups of results (TRECs 8, 2003, and 2008) used in the experiment

Group	Track	Number of queries	MAP of top 10 runs
8	Ad hoc	50	0.3371
2003	Robust	100	0.2697
2008	Blog(opinion)	150	0.4039

These three groups are very different from tracks (ad hoc, robust, and blog), to numbers of queries used (50, 100, 150), to average performances of the selected runs (0.2697, 0.3371, and 0.4039 measured by AP). Such diversified experimental data sets are helpful for us to test the usefulness and adaptivity of the fusion methods involved. In each year group, we choose the top 3, 4, 5,..., up to 10 runs (measured by AP) for the experiment. Each of the selected runs is from a different organization. In this expereiment, we avoid using the multiple runs submitted by the same organization, because those runs are very similar and do not helpful for data fusion methods to improve performance. top systems are selected in order to test if the technique of data fusion can still further improve effectiveness.

Apart from weighted Condorcet voting, CombSum, CombMNZ, the linear combination with performance-level weighting (LCP), MAPFuse [51] (see also Section 5.4.2), the Borda count, and Condorcet are also involved in the experiment. Three different metrics including average precision, recall-level precision, and precision at 10 document level, are used for the evaluation of retrieval performance.

For weighted Condorcet voting, LCP, and MAPFuse, some training data are needed. Therefore, we divide all queries into two groups: odd-numbered queries and even-numbered queries. As before, two-way cross validation is used to test data fusion methods.

In Section 7.2, when discussing how to train weights for weighted Condorcet voting, we assumed that all resultant lists comprise the same group of documents. This is not a realistic assumption. Here we withdraw this assumption and look at a more general situation. The modified approach is: first we set up a pool that comprises a given number (say, 100) of top-ranked documents in all the component results; then all the documents in the pool are divided into two categories: relevant D_r and irrelevant D_i. In the following we only consider those documents that are in the pool and ignore all other documents. For any possible document pairs $< d_a, d_b >$ ($d_a \in D_r$ and $d_b \in D_i$, or $d_b \in D_r$ and $d_a \in D_i$), we check if it is supported by those component results or not. More specifically, for any document pair $< d_a, d_b >$ and component result L_i, if d_a is ranked ahead of d_b in L_i, or d_a appears in L_i but d_b does not, we can see that $< d_a, d_b >$ is supported in L_i and we use '+1' to represent it; if d_b is ranked ahead of d_a in L_i, or d_b appears in L_i but d_a does not, we can see that $< d_a, d_b >$ is not supported in L_i and we use '-1' to represent it; otherwise, neither d_a nor d_b appears in L_i, then it is unclear if $< d_a, d_b >$ is supported or not and we use '0' to represent it. In such a way the table for LDA can be populated. Thus it is able for us to apply LDA on the table to obtain the weights for weighted Condorcet voting.

Example 7.3.1. Let us reconsider Example 7.2.1. $D_r = \{d_1, d_3, d_5\}$, $D_i = \{d_2, d_4\}$, which is the same as in Example 7.2.1. However, this time each information retrieval system only retrieves 3 rather than all 5 documents. $L_1 =< d_1, d_3, d_2 >$, $L_2 =< d_2, d_1, d_3 >$, $L_3 =< d_5, d_4, d_3 >$. Let us see how we can populate the table for LDA.

Some of the pairs for ir_1 (f_1) are used to illustrate the process. One pair is $< d_1, d_2 >$. Both of the documents involved appear in L_1. It can be decided that L_1 supports this pair (+1). Another pair is $< d_1, d_4 >$. d_1 appears in L_1 but d_4 does not. It can be decided that L_1 supports this pair (+1) since we are sure that if retrieved, d_4 must be ranked behind d_1. Another pair is $< d_4, d_1 >$. d_1 appears in L_1 but d_4 does not. For the same reason, we can decide that L_1 does not support this pair (-1). Two other pairs are $< d_4, d_5 >$ and $< d_5, d_4 >$. Both d_4 and d_5 do not appear in L_1, we cannot decide they are supported by L_1 or not. Therefore 0 should be assigned to these two pairs. L_2 and L_3 can be processed similarly. The table we obtain is as follows:

Number	Pair	$f_1(ir_1)$	$f_2(ir_2)$	$f_3(ir_3)$	Category
1	$< d_1, d_2 >$	+1	-1	0	+1
2	$< d_2, d_1 >$	-1	+1	0	-1
3	$< d_1, d_4 >$	+1	+1	-1	+1
4	$< d_4, d_1 >$	-1	-1	+1	-1
5	$< d_3, d_2 >$	+1	-1	+1	+1
6	$< d_2, d_3 >$	-1	+1	-1	-1
7	$< d_3, d_4 >$	+1	+1	-1	+1
8	$< d_4, d_3 >$	-1	-1	+1	-1
9	$< d_5, d_2 >$	-1	-1	+1	+1
10	$< d_2, d_5 >$	+1	+1	-1	-1
11	$< d_5, d_4 >$	0	0	+1	+1
12	$< d_4, d_5 >$	0	0	-1	-1

Now the data in the table are ready for LDA. This time the weights we obtain are 1.273, 0.299, and 1.197 for ir_1. ir_2, and ir_3, respectively. □

7.4 Experimental Results of Fusion Performance

The fusion experimental results are shown in Figures 7.1-7.3 and Tables 7.2-7.4. Figures 7.1-7.3 shows performances (measured by AP) of all data fusion methods

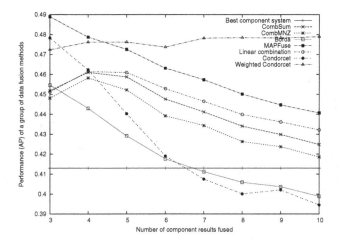

Fig. 7.1 Performance (AP) of several data fusion methods in TREC 8

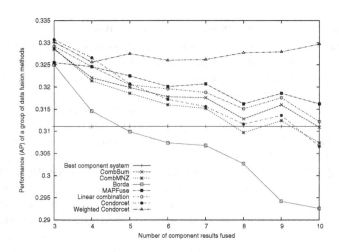

Fig. 7.2 Performance (AP) of several data fusion methods in TREC 2003

in three different data sets. From Figures 7.1-7.3, we can see that the performance of weighted Condorcet is very stable. It is always better than the best component system by a clear margin over all data points and across all three data sets. The linear combination with performance-level weighting performs better than the best system most of the time, but not as good as the best system in TREC 2008 when 8,

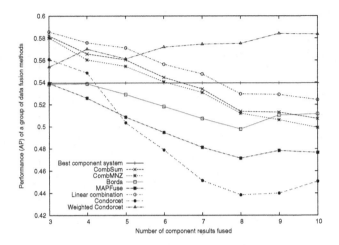

Fig. 7.3 Performance (AP) of several data fusion methods in TREC 2008

9, or 10 systems are fused. MAPFuse performs better than the best system in both
TREC 8 and TREC 2003, but not as good as the best system in TREC 2008. When
a few systems are fused, Condorcet performs better than the best system; but when
more systems are fused, performance of the fused result decreases quickly and not
as good as that of the best system.

Generally speaking, all data fusion methods perform quite differently across different data sets. If we use the ratio of improvement over the best system to evaluate
the fused result, then TREC 8 is the best for all the data fusion methods, TREC 2008
is the worst, while TREC 2003 is in the middle.

Tables 7.2-7.4 shows the average performance (measured by AP, RP, and P@10)
of all the data fusion methods over three data sets. On average, weighted Condorcet
is the best performer among all data fusion methods involved, no matter which metric is used. However, it is always in the second place when three systems are fused. If
measured by average precision or precision at 10 document-level, it is not as good as
Condorcet voting; if measured by recall-level precision, it is not as good as the linear combination with performance-level weighting. Weighted Condorcet performs
better than the best component system by clear margins (8.90% for AP, 5.73% for
RP, and 5.35% for P@10). The linear combination method LCP performs slightly
better than the best component system (4.46% for AP, 2.22% for RP, and 0.70% for
P@10). Among all three data fusion methods that need training data, MAPFuse is
the worst. Its average performance is slightly better than the best component system
if AP or RP is used; but slightly worse than the best component system if P@10
is used for evaluation. If we compare Condorcet voting and weighted Condorcet

Table 7.2 Average performance (AP) of all groups of results in the experiment (WC stands for Weighted Condorcet; the figures in parentheses are the improvement rate of the given data fusion method over the best component system; the figures in bold are the best performed data fusion method in a given experimental setting or on average)

Number	Best	CombSum	Borda	MAPFuse	LCP	Condorcet	WC
3	0.4212	0.4541	0.4393	0.4512	0.4554	**0.4566**	0.4523
4	0.4212	0.4497	0.4321	0.4429	0.4540	0.4458	**0.4573**
5	0.4212	0.4464	0.4227	0.4346	0.4509	0.4215	**0.4550**
6	0.4212	0.4366	0.4145	0.4260	0.4430	0.4050	**0.4572**
7	0.4212	0.4310	0.4085	0.4198	0.4376	0.3915	**0.4597**
8	0.4212	0.4203	0.4022	0.4126	0.4282	0.3834	**0.4605**
9	0.4212	0.4197	0.4028	0.4139	0.4276	0.3865	**0.4635**
10	0.4212	0.4143	0.4010	0.4112	0.4229	0.3840	**0.4640**
Average	0.4212	0.4340	0.4154	0.4265	0.4400	0.4093	**0.4587**
		(3.04%)	(-1.38%)	(1.26%)	(4.46%)	(-2.83%)	(8.90%)

Table 7.3 Average performance (RP) of all groups of results in the experiment (WC stands for Weighted Condorcet; the figures in parentheses are the improvement rate of the given data fusion method over the best component system; the figures in bold are the best performed data fusion method in a given experimental setting or on average)

Number	Best	CombSum	Borda	MAPFuse	LCP	Condorcet	WC
3	0.4377	0.4635	0.4349	0.4618	**0.4627**	0.4590	0.4582
4	0.4377	0.4579	0.4266	0.4555	0.4602	0.4463	**0.4623**
5	0.4377	0.4513	0.4216	0.4482	0.4577	0.4304	**0.4601**
6	0.4377	0.4414	0.4147	0.4420	0.4471	0.4164	**0.4621**
7	0.4377	0.4399	0.4095	0.4358	0.4461	0.4050	**0.4631**
8	0.4377	0.4298	0.4055	0.4299	0.4382	0.3970	**0.4632**
9	0.4377	0.4279	0.4068	0.4305	0.4357	0.4000	**0.4676**
10	0.4377	0.4231	0.4054	0.4281	0.4311	0.3975	**0.4658**
Average	0.4377	0.4419	0.4156	0.4415	0.4474	0.4190	**0.4628**
		(0.96%)	(-5.05%)	(0.87%)	(2.22%)	(-4.27%)	(5.73%)

Table 7.4 Average performance (P@10) of all groups of results in the experiment (WC stands for Weighted Condorcet; the figures in parentheses are the improvement rate of the given data fusion method over the best component system; the figures in bold are the best performed data fusion method in a given experimental setting or on average)

Number	Best	CombSum	Borda	MAPFuse	LCP	Condorcet	WC
3	0.6727	0.6901	0.6833	0.6706	0.6914	**0.7057**	0.7032
4	0.6727	0.6887	0.6714	0.6568	0.6910	0.6843	**0.7094**
5	0.6727	0.6863	0.6673	0.6490	0.6938	0.6644	**0.7094**
6	0.6727	0.6777	0.6582	0.6393	0.6852	0.6444	**0.7041**
7	0.6727	0.6650	0.6488	0.6301	0.6738	0.6272	**0.7067**
8	0.6727	0.6501	0.6429	0.6186	0.6624	0.6122	**0.7079**
9	0.6727	0.6532	0.6431	0.6168	0.6631	0.6146	**0.7128**
10	0.6727	0.6457	0.6399	0.6137	0.6588	0.6123	**0.7162**
Average	0.6727	0.6696	0.6569	0.6369	0.6774	0.6456	**0.7087**
		(-0.46%)	(-2.35%)	(-5.32%)	(0.70%)	(-4.03%)	(5.35%)

voting, then the improvement rate of the latter to the former is 12.07% for AP, 10.45% for RP, and 9.77% for P@10. Both Condorcet voting and Borda count do not rely on scores and training data. If we compare them, then we can see that the average performance of Condorcet fusion is slightly worse than that of Borda count, though the Condorcet voting is better than the Borda count when 3 or 4 systems are fused.

If we use both metrics AP and RP, then on average all data fusion methods excluding the Borda count and Condorcet voting are better than the best component system. But if we use P@10, then only weighted Condorcet voting and the linear combination method with performance-level weighting are better than the best component system, while the others are worse.

Finally, we can notice one phenomenon: when more and more information retrieval systems are used for fusion, weighted Condorcet voting is the only one that becomes better and better (only slightly though), while all others become worse and worse. This, of course, does not necessarily indicate that the number of component systems is a negative factor that affects the performance of these data fusion methods. Recall that the component systems are added in the order of decreasing performance. Therefore, when more and more component systems are added, the average performance of all component systems becomes lower and lower, and the deviation of each component system's performance from the average performance becomes larger and larger. This can explain why almost all data fusion methods, especially those treat all component systems equally, become worse off when more systems are fused.

7.5   Positive Evidence in Support of the Hypothesis

At the end of Section 7.1, we proposed a hypothesis. It claims that unfavourable conditions do more damage to Condorcet than to some other methods such as CombSum and the Borda count. In this section, we would provide some evidence to support this hypothesis. We reuse the same experimental data and results reported in Section 7.3.

First we compare the performance of data fusion methods when different number of component systems are fused. To make the comparison straightforward, the performance of the fused result of three component systems is used as baseline, then we observe how the performance deteriorates when more component systems are involved. Figures 7.4 and 7.5 present the results, in which AP and P@10 are used as the metric, respectively. From Figures 7.4 and 7.5, we can see that all data fusion methods become worse when more component systems are involved. This is reasonable because it is in line with the average performance of all component systems.

The interesting part is the different deteriorating rates of the three data fusion methods involved. In both Figures 7.4 and 7.5, we can see that Condorcet is the

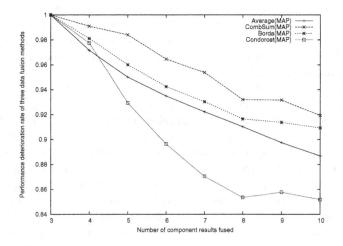

Fig. 7.4 Performance comparison (AP) of several data fusion methods

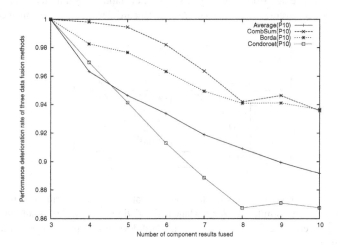

Fig. 7.5 Performance comparison (P@10) of several data fusion methods

most affected when more component systems are fused. As a matter of fact, when 5 or more component systems are fused, the deteriorating rate of it is even bigger than that of the average performance of all component systems; while for the two other methods, their deteriorating rates are not as big as that of the average performance. This interesting phenomenon is a piece of positive evidence in support of the hypothesis proposed in Section 7.1.

Next we use the multiple linear regression technique to explore the relation between fusion performance and some statistics of component systems. We use average similarity (ave_s) and standard deviation of similarity (dev_s) of all component systems, average performance (ave_p) and standard deviation of performance (dev_p) of all component systems, and the logarithm of number of systems ($\ln(num)$), as independent variables, and fusion performance (f_ave_p) as the dependent variable, to conduct the multiple linear regression. In order to obtain ave_s, we calculate the average Spearman's ranking coefficient of two component results across a group of queries. Then ave_s is defined as the average of that for all possible pairs of component results. For both ave_p and f_ave_p, average precision is used as the metric for performance evaluation. The linear regression results are shown in Tables 5-7, for Condorcet voting, the Borda count, and CombSum, respectively.

Table 7.5 Linear regression results of the Condorcet voting ($\ln(num)$: logarithm of number of systems; ave_s: average similarity; dev_s: standard deviation of similarity of all component systems; ave_p: average performance; dev_p: standard deviation of performance of all component systems)

Model	Coefficients		t	Significance level
	Value	Std. Error		
$\ln(num)$	.043	.008	5.215	.000
ave_s	-.141	.029	-4.914	.000
dev_s	-.226	.058	-3.891	.000
ave_p	1.026	.016	65.487	.000
dev_p	.104	.050	2.065	.039
R^2	.688			

Table 7.6 Linear regression results of the Borda count ($\ln(num)$: logarithm of number of systems; ave_s: average similarity; dev_s: standard deviation of similarity of all component systems; ave_p: average performance; dev_p: standard deviation of performance of all component systems)

Model	Coefficients		t	Significance level
	Value	Std. Error		
$\ln(num)$	.007	.004	1.626	.104
ave_s	-.112	.015	-7.588	.000
dev_s	.323	.030	10.831	.000
ave_p	1.101	.008	136.685	.000
dev_p	.168	.026	6.500	.000
R^2	.911			

Table 7.7 Linear regression results of CombSum (ln(*num*): logarithm of number of systems; *ave_s*: average similarity; *dev_s*: standard deviation of similarity of all component systems; *ave_p*: average performance; *dev_p*: standard deviation of performance of all component systems)

Model	Coefficients		t	Significance level
	Value	Std. Error		
ln(*num*)	-.004	.003	-1.098	.272
ave_s	-.105	.011	-9.265	.000
dev_s	.264	.023	11.540	.000
ave_p	1.128	.006	182.622	.000
dev_p	.411	.020	20.695	.000
R^2	.951			

In Tables 7.5-7.7, a significance level of t means that the p-value is t; a significance level of .000 means that the p-value $< .0005$. In overall, the model for CombSum is the most accurate, with a R^2 value of .951; which is followed by the model for the Borda count, with a R^2 value of .911; while the model for Condorcet voting is the least accurate, with a R^2 value of .688. This demonstrates that both the Borda count and CombSum can be described with a linear model very accurately and Condorcet voting is more complex and more difficult to describe with a linear model than the Borda count and CombSum. However, the model for Condorcet is still quite accurate. There are certain similarity among these three methods. *ave_p* is always the most significant factor that affect the performance of all three data fusion methods. In all three methods, the coefficient values of it (1.026, 1.101, and 1.128) are always much bigger than others. *dev_p* is always a significant factor that makes positive contribution to fusion performance, while *ave_s* is always a significant factor that makes negative contribution to fusion performance. It can also be observed that the Borda count and CombSum are more similar to each other than either of them to Condordet fusion. In Condorcet voting, *dev_s* makes negative contribution, while in both Borda count and CombSum, *dev_s* makes positive contribution to fusion performance. This is a major difference between Condorcet voting and the other two. This confirms that uneven similarity between component results, indicated by standard derivation of similarity, is especially harmful to Condorcet voting, but neither the Borda count nor CombSum. On the other hand, if we compare the coefficient values of *dev_s*, then we find that the one in Condorcet (.104) is smaller than its peers in the Borda count and CombSum (.168 and .411). Therefore, in order to achieve good performance, weighted Condorcet voting needs to be used in more situations than some other fusion methods such as weighted Borda count and the linear combination method.

Chapter 8
Fusing Results from Overlapping Databases

With the rapid development of the Internet and WWW, numerous on-line resources are available due to efforts of research groups, companies, government agencies, and individuals all over the globe. How to provide an effective and efficient solution to access such a huge collection of resources for end users is a demanding issue, which is the major goal of federated search, also known as distributed information retrieval. In this section, we are going to discuss the two key issues involved, i.e., resource selection and especially result merging, with the assumption that partial overlap exists among different resources. In a sense, merging results from overlapping databases looks somewhat like the data fusion problem, in which results are obtained from identical databases.

8.1 Introduction to Federated Search

Federated search allows the simultaneous search of multiple searchable resources. A user makes a single query request which is distributed to the search engines participating in the federation. The federated search system (broker) then merges the results that are received from the search engines (database servers) and presents them to the user. Usually, the process of a federated searching consists of the following three steps:

- Transforming a query and broadcasting it to a group of participating databases.
- Merging the results collected from the database servers.
- Presenting them in a unified format to the user.

Figure 8.1 shows the architecture of a federated search system with n ($n \geq 2$) search engines. From the technical point of view, there are two key issues: resource selection (also known as database selection) and result merging. If a large number of resources participate in the federation, then it might not be a good option to broadcast any query to all the participants. Selecting a subset of most useful resources can

S. Wu: Data Fusion in Information Retrieval, ALO 13, pp. 149–180.
springerlink.com   © Springer-Verlag Berlin Heidelberg 2012

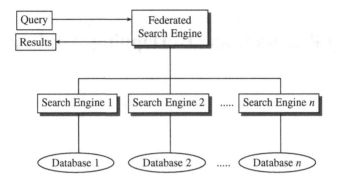

Fig. 8.1 The architecture of a federated search system with n search engine components

be more cost-effective. After receiving the results from multiple resources, result merging is necessary for improving the usability and performance of the system.

Usually, researchers assume that there is no or very little overlap across different resources. If all participating resources are co-operative, then it is possible to obtain useful statistics for resource selection. GlOSS/gGlOSS [37], CVV [129], CORI [17] are some proposed methods in such a scenario.

If some participating databases are not co-operative, more effort is needed in order to obtain some necessary information. One possible solution is to use a centralized sample database which comprises sampled documents from all participating resources. Then the performance of every resource is estimated by the sample database in which documents are served as representatives of each resource. Several resource selection methods, such as ReDDE [86], CRCS [81], and SUSHI [93], are based on a centralized sample database. Another possible solution is to carry out lightweight probe queries [38] so as to find useful information for selecting a group of resources.

Fuhr [36] introduced a decision framework for resource selection. Apart from number of relevant documents, several other aspects such as response time and monetary costs are also considered.

As to the result merging part, Round-robin is a simple merging method, which takes one document in turn from each available resultant list. However, the effectiveness of such a method depends on the performances of component database servers and the contents of component databases. If all the results are of similar quality, then Round-robin performs well; if some results are much worse than the others, then Round-robin becomes very poor.

Voorhees, Gupta and Johnson-Laird [100] demonstrated a way of improving the above-mentioned Round-robin method. By running some training queries, they can estimate the performance of each resource. When a new query is encountered, a weighted Round-robin can be used. Rather than taking one document each time from one database, it takes a certain number of documents from one database according to the estimated performance of it.

Callan, Lu and Croft [16, 17] proposed a merging strategy based on the scores achieved by both resource and document. The score for each resource is calculated by a resource selection algorithm, CORI, which indicates the "goodness" of a resource with respect to a given query among a set of available resources. The score of a document is the value that the document obtains from the resource, which indicates the "goodness" of that document to the given query among a set of retrieved documents from that resource. Yuwono and Lee [129] proposed another method for converting local document scores to global scores.

SSL [85, 87] is a semi-supervised, sample-based learning method that trains a regression model for each database that maps document scores into their global scores. One requirement for the model to work is that some overlapping documents exist in multiple databases. Another sample-based result merging method is proposed by Shokouhi and Zobel [84]. More result merging methods are investigated in [73, 102, 127] among others. There is also a recent survey paper on federated search [83].

In [56], three typical cases are identified based on the relationship between the databases that are associated with participating search engines :

- The databases are pair-wise disjoint or nearly disjoint;
- The databases overlap but are not identical;
- The databases are identical.

Cases 1 and 2 may occur quite often in a federated search environment. Case 3 is an ideal situation and usually does not occur in a typical federated search environment. However, it has been exploited to improve retrieval effectiveness known as the data fusion technique. Partial overlaps happen in many situations. For example, there are dozens of general-purpose web search engines, their document databases overlap heavily with each other. For a portal or meta-search engine to retrieve documents from them, overlapping should be considered in the result merging for better performance. Another example concerns the scientific literature. Every year scientists and researchers publish a huge number of papers in various periodicals and conferences. Many of these publications are available on-line through digital libraries and/or electronic information services. Considering computer science-related databases, there are dozens of them available. E.g., ACM digital library, Applied Science and Technology Plus, Compendex, Computer and Information Systems Abstracts, CSA, Current Contents Connect, Scientific & Technical Information Network, Electronics and Communications Abstracts, IEEE/IEE Electronic Library Online, Science Citation Index, Web of Knowledge and many others. Considerable overlaps exist among these databases. To provide a service that can search many of them automatically at the same time is definitely an advantage for use. On the other hand, a solution that is ignorant of overlaps among databases may degrade the performance of retrieval considerably. Therefore, the effect of overlapping should be investigated in several activities such as resource selection and result merging in federated search.

8.2   Resource Selection

In the following discussion, resource relevance, monetary costs, response time, and overlapping rates among databases are going to be considered by a multi-objective optimization model [114].

8.2.1   A Basic Resource Selection Model

The **Utility Function Method** [72] could be used to deal with this multi-objective optimization problem. First, we define a utility function for each of the four objectives depending on their importance, then a total utility function can be defined. In our case, we adopt the common linear function by defining a coefficient for each of the objectives. A total utility function could be defined as follows:

$$U = k_1 R_{res} + k_2 T_{res} + k_3 C_{res} + k_4 DR_{res} \tag{8.1}$$

where $R_{res}, T_{res}, C_{res}$, and DR_{res} are average measures for the four objectives at which we are aiming: relevance, time, cost, and duplicates ratio. For any query submitted, the federated search system (broker) has to make a decision about which database server to query, and how many documents should be fetched from each selected server. Usually the options are numerous. For each option, R_{res}, T_{res}, C_{res}, and DR_{res} will be given a corresponding value according to the formulae which we will discuss later in this section. All the parameters are required to be normalized (ranged in [0,1]), thus we can treat all factors equally. k_1, k_2, k_3, and k_4 are coefficients, $k_1 \geq 0, k_2 \leq 0, k_3 \leq 0, k_4 \leq 0$. The values of these coefficients can be adjusted for different policies. For example, user A, concerned more with relevance and less with costs and time, could set $k_1 = 1$, $k_2 = k_3 = 0$, and $k_4 = -1$; while a setting of $k_1 = 0.5$, $k_2 = k_3 = -1$, and $k_4 = -1$ is suitable for user B, who intends to get a few relevant documents cheaply and quickly. In many cases document duplicates are annoying so we could set -1 as a default value for k_4. After setting up all the coefficients, the next question is how to maximize it.

Assuming that for any query, the user always specifies a number n instructing the federated search system to return the top n estimated relevant documents. It is the federated search system's responsibility to work out a solution that can best meet the requirements of the user.

Before going into details of parameter estimation, let us introduce some notation that will be used in this section. Suppose there are m available database servers in all for consideration, DBs denotes the set of database servers we have, and D_i ($1 \leq i \leq m$) denotes the i-th database server among them. In the following, all the parameters are related to the result of a given query q. L_i denotes the result set retrieved from D_i with respect to q. x_i denotes the number of documents in L_i. L_{ij} denotes the j-th document of L_i, and D_S_i denotes the "goodness" score of D_i. T_All_{ij} denotes the time spent for obtaining the first j documents in L_i, while C_{ij} denotes the costs

Table 8.1 Symbols and their meanings used in Section 8.2

Symbol	Description
DBs	resource (database) set.
D_i	the i-th resource in DBs.
m	number of resources in DBs.
m'	number of resources whose documents are in the result.
q	the given query.
n	number of documents specified by the user for query q.
L_i	document set obtained in D_i when performing q.
L_{ij}	the j-th document in L_i.
x_i	number of documents required from D_i or number of documents in L_i.
C_{ij}	charge for obtaining L_{ij}.
T_All_{ij}	time needed for the first j-th documents in L_i.
Avg_T_{ij}	average time for each of the first j documents of L_i.
NAT_{ij}	normalized form of Avg_T_{ij}.
Avg_C_{ij}	average charge for each of the first j documents of L_i.
NAC_{ij}	normalized form of Avg_C_{ij}.
D_S_i	relevance score of D_i.
Avg_S_i	average score of D_i for each document in L_i.
NAS_i	normalized Avg_S_i.
DR_{ik}	document duplicates ratio of L_i and L_k.
U	utility function.

(charge) that the user needs to pay for $L_{i,j}$. SR_{ik} denotes the duplicates ratio among the two sets of documents L_i and L_k.

As a summary, Table 8.1 shows all the symbols used in this section with their meanings.

For the application of this model, one of the key requirements is to assign appropriate values to the parameters in the model. First let us look at how to obtain relevance scores. In [36], Fuhr proposed a probabilistic method to estimate the number of relevant documents to a query in a resource. A statistical method for the same purpose was proposed in [53]. Some other selection methods (e.g., [37], [129], and [17]) calculated scores according to their own formulae for all available resources. Any reasonable resource selection algorithms such as [17], [37], and [129] can be used to calculate scores. For simplifying the discussion, we assume that the same selection algorithm is used for all the resources involved. Suppose D_i $(1 \leq i \leq m)$ is assigned a score of D_S_i and is required to provide x_i $(\sum_{i=1}^{m} x_i = n)$ documents, then we define D_{max} as:

$$D_{max} = \max_{(1 \leq i \leq m)} D\_S_i$$

and define x_{max} as:

$$x_{max} = \max_{(1 \leq i \leq m)} x_i$$

Based on that, we have:

$$Avg\_S_i = \frac{D\_S_i}{D_{max}}$$

$$NAS_i = \frac{Avg\_S_i}{x_{max}}$$

$$R_{res} = \frac{1}{n} \sum_{i=1}^{m} NAS_i * x_i$$

In such a way, R_{res} is normalized and takes value in [0, 1]. In case of different algorithms used for different resources, a mapping is needed for making these scores comparable.

Fuhr's or Liu, et al.'s methods may also be used here. First, we estimate the number of all relevant (useful) documents n_i in resource D_i to query q. Then we could use the same method above, simply taking the number of relevant (useful) documents as its score.

The time needed for retrieving documents from a database server can be divided into two major parts: computation time and transmission time. The former is the time used in the database server to find the documents, while the latter is the time used to transmit the documents from the database server to the broker. It is usually the case that the number of documents we fetch from the database server has little impact on the computation time [36], but that number has a directly proportional effect on the transmission time. Suppose for resource D_i, T_{i1}, T_{i2},... is the time needed to fetch the first, second,... documents. We can expect that T_{i1} is a relatively long time period that covers most of the computation time of that query, while T_{i2} and its successors are mainly composed of transmission time. The time can be estimated either by carrying out some sample queries or from past queries.

Suppose the value of T_All_{ij}, the time needed for fetching the first j documents, is known for any j in L_i, we use it to calculate T_{res} in the following way. First, we calculate T_{ij} and Avg_T_{ij} by

$$T_{ij} = T\_All_{ij} - T\_All_{i,j-1}$$

and

$$Avg\_T_{ij} = \frac{T\_All_{ij}}{j}$$

Then we define:

$$T_{max} = \max_{(1 \leq i \leq m, 1 \leq j \leq n)} T_{ij}$$

and

$$NAT_{ij} = \frac{Avg\_T_{ij}}{T_{max}}$$

Thus, T_{res} can be defined as:

$$T_{res} = \frac{1}{n} \sum_{i=1}^{m} NAT_{ix_i} * x_i$$

A similar approach can be used for monetary cost. Suppose we know C_{ij}, which denotes the charge that the user needs to pay for L_{ij}. We define

$$Avg\_C_{ij} = \frac{1}{j} \sum_{k=1}^{j} C_{ik}$$

and

$$C_{max} = \max_{(1 \leq i \leq m \wedge 1 \leq j \leq n_i)} C_{ij}$$

which yields:

$$NAC_{ij} = \frac{Avg\_C_{ij}}{C_{max}}$$

Finally, we have:

$$C_{res} = \frac{1}{n} \sum_{i=1}^{m} NAC_{ix_i} * x_i$$

Finally, the presence of document duplicates may degrade greatly the effectiveness of the broker. In the worst case, several servers may just returns the same lists of documents. In order to deal with duplicated documents in a resource selection model, specific information needs to be stored in the broker. Ideally, for every query and a certain number of documents, storing the estimated value of duplicates that exists between every pair of resources. However, it is not a practical solution by any means. In the following, we discuss a few possible solutions that can be used for this. In all cases, we assume that that every resource provides the same number of documents x.

- Setting up DR_{ij} manually. In some cases, the task is straightforward. For example, for a web site and any of its mirrors, DR_{ij} is always equal to 1. An opposite situation is that DR_{ij} is equal to 0. That is the case for resources that cover totally different fields, or resources that cover a similar field, but no intersection is expected. For example, the ACM digital library and the AMA (American Medical Association) data services and publications may fall in this category.
- Server-pair general ratio. This method assumes that for any query, duplicates ratio between documents retrieved from two different servers are always the same. By sending a predefined group of sample queries to two servers, we can identify the number of duplicates from a given number of top retrieved documents and average the ratios over different queries. After that, the figure can be used in every case regardless of the query involved.
- Server-pair clustered ratio. This method divides all the documents in two resources into a common set of clusters. We estimate the ratio for each cluster by sending some sample queries to the two servers. When a new query is submitted by a user, we select a cluster that is most relevant to the query, and use the duplicates ratio of that cluster for the query.
- Past queries-based method. Suppose the broker has a log mechanism. For every query processed, we compare the results of different servers and record each

duplicates ratio into the log. When a new query is submitted, the broker finds a similar query in the log and uses the ratio of that query for the new query.

The last three methods can be implemented automatically in the broker. Based on the above methods, we can estimate the number of document duplicates of every pair of database servers for any query. Then DR_{res} can be defined as:

$$DR_{res} = \frac{1}{m'-1} \sum_{i=1}^{m'-1} \{Const \sum_{j=i+1}^{m'} (DR_{ij} * min(x_i, x_j))\} \qquad (8.2)$$

where $Const = \frac{1}{n - \sum_{k=1}^{i} x_k}$. There are $\frac{m' * (m'-1)}{2}$ different pairs for m' resources. We divide them into $m' - 1$ groups. The i-th ($i = 1, 2, \ldots m' - 1$) group includes all those pairs (D_i, D_{i+1}), $(D_i, D_{i+2}),\ldots,(D_i, D_{m'})$. Therefore, the total number of duplicates in group i is calculated by $\sum_{j=i+1}^{m'}(DR_{ij} * min(x_i, x_j))$. Because $x_1 + x_2 + \ldots + x_{m'} = n$, the maximum number of duplicates in this group is $n - \sum_{k=1}^{i} x_k$. In Equation 8.2, $Const$ is used as the normalizing factor for every group. When $x_1 = x_2 = \ldots = x_{m'} = n/m'$, and $DR_{ij} = 1$ for any i and j, $DR_{res}=1$.

8.2.2   Solution to the Basic Resource Selection Model

Given the framework presented in Section 8.2.1, now the task is to maximise U. In the following, we assume that for each resource D_i, the size of L_i is always n. This could be done in the following way: if the size of the retrieved document set for D_i is greater than n, it is sufficient to keep the first n documents and discard the rest; if the size of the document set is less than n, we can add some "dummy" documents with very low relevance but high time and monetary cost. In such a way, we can guarantee that these dummies will not be selected. The only constraint is that we need to have at least n documents in all. Otherwise, it is unavoidable for the algorithm to pick up some dummy documents. However, if that is the case, we do not need to run the algorithm in the first place.

One way of obtaining the overall optimum solution is to enumerate all possible candidates and decide which one is the best. Note that n documents in m resources can be mapped into a m-digit number in a base $n + 1$ numerical system. The problem can be mapped into the question that finding out all such numbers whose digits in all places sum to n. The following rules are always true for the numbers which satisfies our requirement:

- $\underbrace{0\ldots\ldots0}_{m-1}n$ is the smallest;

- $n\underbrace{0\ldots\ldots0}_{m-1}$ is the largest;

- if T is a satisfied number whose digit in the units place is not 0, then beyond T, $T + n$ is the smallest one that satisfies our requirement.

- if T is a satisfied number whose lower $l+1$ places having a form of n' $\underbrace{0........0}_{l\wedge1\leq l<m-2}$,

  then $T + (n - n' + 1)\underbrace{0.........0}_{l-1}(n' - 1)$ is the next one beyond T.

Based on the above observation, Algorithm 8.1 shown in Figure 8.2 lists every so-lution. For each of them, Algorithm 8.1 calls Procedure CalUtility to calculate its utility. Algorithm 8.1 compares all the utility values obtained and keeps the best as the final solution.

Algorithm 8.1

```
01      Input: m, n; //There are m resources and n documents are required
02      Output: best_utility, x(1..m); // For the optimum solution
03      //Variable used: p points to the lowest non-zero place
04      //Variable used: A(0..m) for keeping a n-digit number
05      best_utility := -∞; //Lines 5-8 for initialization
06      for i := 0 to m − 1 do A(i) := 0;
07      A(m) := n; //The first solution
08      while (A(0) ≠ 1) do
09      { utility := CalUtility(A);
10          if (utility > best_utility)
11          { best_utility := utility;
12              for j := 1 to m do X(j) := A(j);
13          }
14          //Lines 15-25 for setting up next solution
15          if (A(m) ≠ 0) //the digit in the units place is not 0
16          {
17              A(m)−−; A(m-1)++;
18              p := m-1;
19              if (a(p)=n+1) {a(p) := 0; a(p-1)++; p−−; }
20          }
21          else // the digit in the units is 0
22          {p := m;
23              while (a(i=0) p−−;
24              A(m) := A(p)-1; A(p) := 0; A(p-1)++;
25              p := p-1;
26              if (a(p)=n+1) {a(p) := 0; a(p-1)++; p−−; }
27          }
28      }
29
30      Procedure CalUtility(A)
31      //Use Equation 8.1 to calculate the utility of a given solution
```

Fig. 8.2 Calculating the optimum solution for the basic multi-objective model

The Procedure CalUtility works as follows. Suppose that the values of m, n, k_1, k_2, k_3, k_4, D_S_i, C_{ij}, T_All_{ij}, and DR_{ij} for ($1 \leq i \leq m$, $1 \leq j \leq n$) are known. For a solution $X = \{x_1, x_2, ..., x_m\}$, we can calculate its utility directly by Equation 8.1.

First, we can calculate NAT_{ij}, NAC_{ij}, and NAS_i for any $(1 \leq i \leq m, 1 \leq j \leq n)$ from given conditions, which takes $O(mn)$ time, then we can calculate R_{res}, T_{res}, and C_{res} in $O(m)$ time. For the duplicates ratio, it takes $O(m^2)$ time. The final step is to sum them together.

For m resources and n documents, the number of solutions that Algorithm 8.2.1 generates is given and proved by the following Theorem.

Theorem 8.2.1. For m resources and n documents, the number of all possible solutions is:

$$A(m,n) = \frac{m(m+1)....(m+n-1)}{n!}$$

Proof: we prove it by mathematical induction.

Basis: for any number $i = 1,2,...n$, $A(1, i) = 1$. That means if we have just one resource, there is just one way for us to handle all the documents.

Assumption: for any number i $(1 \leq i \leq n)$,

$$A(m,i) = \frac{m(m+1)...(m+i-1)}{i!}$$

Induction: we need to prove

$$A(m+1,i) = \frac{(m+1)(m+2)...(m+i)}{i!}$$

As a matter of fact, $A(m+1, i)$ means we have $m+1$ resources and i documents. We can put 0, or 1, or 2, ...,or i documents into the first resource, that will leave i, or $i-1$, or $i-2$,..., or 0 documents for the rest m resources. Thus,

$$
\begin{aligned}
A(m+1,i) &= A(m,i) + A(m,i-1) + A(m,i-2) + ... + A(m,1) + 1 \\
&= \frac{m(m+1)...(m+i-1)}{i!} + \frac{m(m+1)...(m+i-2)}{(i-1)!} + ... + m + 1 \\
&= \frac{m(m+1)...(m+i-1)}{i!} + \frac{m(m+1)...(m+i-2)i}{i!} + ... + \frac{mi!}{i!} + \frac{i!}{i!} \\
&= \frac{m...(m+i-1) + m...(m+i-2)i + ... + mi! + i!}{i!}
\end{aligned}
\tag{8.3}
$$

Then, we can decompose A(m+1, i) in Assumption into many items step by step, each time split the last factor:

$$
\begin{aligned}
A(m+1,i) &= \frac{(m+1)(m+2)...(m+i)}{i!} \\
&= \frac{(m+1)...(m+i-1)m + (m+1)(m+2)...(m+i-1)i}{i!} \\
&= ... \\
&= \frac{m...(m+i-1) + m...(m+i-2)i + ... + mi! + i!}{i!}
\end{aligned}
\tag{8.4}
$$

Comparing Equations 8.3 with 8.4, we can see that they are equal.   □

According to Stirling's formula, $n! \approx (n/e)^n \sqrt{2\pi n}$. Therefore, we have

$$A(m,n) = \frac{(m+n-1)!}{n!(m-1)!}$$

$$\approx \frac{((m+n-1)/e)^{m+n-1}\sqrt{2\pi(m+n-1)}}{(n/e)^n\sqrt{2\pi n}((m-1)/e)^{m-1}\sqrt{2\pi(m-1)}}$$

$$\approx \frac{(m+n)^{m+n}}{n^n m^m}$$

When $m = n$, the above function gets its maximum, therefore the worst time complexity of $A(m,n)$ and the worst complexity of Algorithm 8.1 is $O(2^{m+n})$.

Algorithm 8.1 can be further improved. The utility in Equation 8.1 for a given solution can be divided into four parts. The first three parts have a common property. That is, for any given resource, the utility of a solution from one part is only determined by the number j of documents that is involved in the result. For example, for relevance it is $j * NAS_i$, if the first j documents in resource D_i are involved in the solution. But for document duplicates ratio, the situation is different, since that part of the utility is decided by all the resources that have some documents appearing in the resultant list. If we pre-calculate the value of the first three parts of U for every resource i and every possible number j of documents:

$$U'_{ij} = j * (k_1 NAS_i + k_2 NAT_{ij} + k_3 NAC_{ij}) \tag{8.5}$$

then, for every solution, we can get its partial utility value by one scan of each x_i value of every resource D_i, that is $\sum_{i=1}^{m} U'_{(i,x_i)}/n$. However, there is no simple way of obtaining DR_{res}, which has to be handled as described before. In such a way, procedure CalUtility needs $O(m^2)$ time for every execution.

8.2.3 Some Variants of the Basic Model and Related Solutions

The basic model in Section 8.2.1 reflects precisely the multi-objective resource selection problem, but the solution given in Section 8.2.2 is not efficient enough for practical use. This is due to the objective of document duplicates. In this subsection we will remove some of the requirements to improve efficiency. These variants of the basic model are still sound but are more efficient, as the evaluation reported in Section 8.2.4 will show.

a. The Static Variation Model
One possible way is to define an average document duplicate ratio between resource D_i and all other available resources as:

$$Avg\_DR_i = \frac{1}{m-1} \sum_{k=1 \wedge k \neq i}^{m} DR_{ik}$$

Then we can define DR_{res} as:

$$DR_{res} = \frac{1}{n} \sum_{i=1}^{m} x_i * Avg\_DR_i \tag{8.6}$$

Comparing the above equation (Equation 8.6) with Equation 8.2, we can see that in Equation 8.6, each addend can be calculated out in a more efficient way than before. In this variant of the basic model, all other three parts in the utility function are the same. Since the utility that each resource earns is calculated independently from other resource, we call this model the static variation model.

Algorithm 8.2

```
01      Input: m, n, U(1..m, 0..n);
02      Output: B(1:m); // B(1:m) is for the optimum solution
03      for i := 1 to m do
04      for j := 1 to n do
05      W(i,j) := 0;
06      for i := 0 to n do X(i) := U(1,i);
07      for i := 2 to m do
08      for j := 1 to n do
09      for k := 0 to j − 1 do
10      {
11          if ((X(j) < (X(k) + U(i,j-k)))
12          {(X(j) = (X(k) + U(i,j-k)); W(i,j) := k;}
13      }
14      t := n;
15      for i:=m to 1 do
16      {
17          B(i) := W(i,t); t := t - W(i,t);
18          if (t=0) B(i) := 0;
19      }
```

Fig. 8.3 Calculting the optimum solution for the static variation model

Now that we make our decision by only considering the Avg_DR_i value of any given resource, such a model is less accurate than the basic model. However, we can benefit from this simplified model since Algorithm 8.1 can now be implemented more efficiently. We can evaluate the matrix U_{ij} using the following equation:

$$U_{ij} = (k_1 NAS_i + k_2 NAT_{ij} + k_3 NAC_{ij} + k_4 Avg\_DR_i) * j \tag{8.7}$$

In such a situation, the "divide-and-conquer" algorithm (Algorithm 8.2) described in Figure 8.3 could be used. The dividing step iterates over the number of resources. Since the utility for documents in each resource is not related to any other resource that contributes to the result, we can use the maximum computed for the i-th resource in computing the maximum for the $(i+1)$-th resource. The time complexity of Algorithm 8.2 is $O(mn^2)$. This algorithm gives optimum solution provided that all parameters are accurate and U_{ij} in Equation 8.7 is monotonic.

b. A Greedy Algorithm

Suppose only one maximum exists in the utility function for each resource, then we can use a greedy policy to speed up the algorithm for the static variation model. In practical situations, the above monotonic assumption of utility does not always hold. However, relaxing the above restriction is rather straightforward. If the utility function for a given resource has several local maximum points, we can always take any one of them as the global one. The greedy algorithm, or Algorithm 8.3, is described in Figure 8.4.

Algorithm 8.3

```
01      Input: m, n, opt(1..m), X(1..m), U(1..m, 1..n)
02      Output: res_n, X(1..res_n);
03      for i := 1 to m do F(i) := U(opt(i), x(i));
04      count := X(1); if (count ≥ n) {res_n := 1; X(i) := n; exit;}
05      for i:= 2 to m do
06      {
07         for j := 1 to i-1 do
08         {
09            while (F(i) < F(j));
10            {
11               x(i)++;
12               F(i) = U(opt(i),x(i));
13               count++;
14               if ( count = n) res_n := i-1; exit;
15            }
16         }
17         count := count + F(i);
18         if (count ≥ n) { res_n := i; X(i) := X(i)-(count-n) exit;}
19      }
```

Fig. 8.4 Computing a solution by a greedy policy

The basic idea of the greedy algorithm is to find the maximum point x_i for each D_i, then rank them by their utility value in descending order. Note that the solution aims at selecting a group of resources for retrieving n documents in all. Two sets can be used. One includes all the available resources and the other is the solution set. At first, the solution set is empty. The algorithm repeatedly removes the highest ranked D_i from the available resource set, put it into the solution set, with its x_i documents added, until the solution reaches n documents. In addition, before adding a new resource into the solution set, we try to increase the number of documents of every resource already in the result as long as its utility is no less than the newcomer's. Finally, if we cannot have just n documents, we reduce the document number of the last incomer to fit n. The algorithm is shown in Figure 8.4.

In Figure 8.4, $opt(1..m)$ gives the subscripts of all resources sorted by their maximum of utility function in descending order, while $X(1..m)$ initially gives the numbers of documents in resources when they reach their maximum values. When the algorithm ends, the variable res_n gives the total number of resources whose documents are selected in the result, in gives the total number of documents involved in the result, and $X(1..res\_n)$ gives the number of documents each resource contributes. This algorithm requires at most $O(m)$ to get one document. Since we need n documents in all, the worst time complexity of the algorithm is $O(mn)$.

The Dynamic Variation Model
Algorithm 8.3 selects resources one by one according to their utility function performance. This provides us a possible way to improve the algorithm on document duplicates.

Algorithm 8.4

```
01      Input: m, n, X(i), U'(1..m, 1..n);
02      Output: X(1..m), D_R;
03      D_R := { }; D_U := D; //D_R denotes the resources in the result already
04      select D_i from D_U which has the maximum utility value U(i, j);
05      D_R := { D_i }; D_U := D_U - { D_i };
06      count := X_i;
07      if (count ≥ n) then exit;
08      while (1=1)
09      { select D_i from D_U that maximizes Equation 8.8;
10          for each D'_i in D_R,
11          if (U'(i,X(i)) < U'(i',x(i'))),
12          take more documents from D_i' and change the values of count and X(i) accordingly;
13          if (count ≥ n) exit;
14          D_R := D_R + { D_i }; D_U := D_U -{ D_i };
15          count := count + j;
16          if (count ≥ n) then exit;
17      }
```

Fig. 8.5 Computes a solution for the dynamic variation model

Suppose we are running that algorithm, having m' resources $\{D_1, D_2, \ldots, D_{m'}\}$ in the resultant set already, but the total number of documents is still less than the required number (n). In such a situation, we need to select some more documents from other resources. One way to select a new resource is to consider the overlap between the candidate and those resources that having been included in the result, instead of all available resources. This will be more accurate for evaluating duplicated document ratios of the result. The next D_i we select is the one that gets the maximum value in the following equation:

$$U_n = max_{D_i \in D\_U} \left(U'_{ix_i} + \frac{1}{\sum_{j=1}^{m'} x_j} \sum_{j=1}^{m'} (DR_{ij} * min(x_i, x_j)) \right) \tag{8.8}$$

where D_U is the set of the available resources that have not been involved in the result, x_i is the number of documents in L_i, the number at which D_i gets its maximum of the utility function, and m' is the number of resources already in the result. U'_{ij} is defined in Equation 8.5. In contrast to the static variation model, we call this the dynamic variation model. Likewise Algorithm 8.3, we can design a greedy algorithm for such a model. Before adding documents of a new resource into the result, we expand the number of documents of those resources already in the result as long as it is profitable. One more thing we should consider is the first resource to be selected. Since there is no resource in the result, the process is exactly the same as in Algorithm 8.3 and the aspect of database overlaps do not need to be considered. Algorithm 8.4 is described in Figure 8.5.

In Algorithm 8.4, line 4 can be done in $O(mn)$ and the loop between lines 8 and 17 can be done in $O(mn)$ as in Algorithm 8.3. Therefore, the total complexity of the algorithm is $O(mn)$.

8.2.4 Experiments and Experimental Results

Two experiments were conducted to evaluate the performances of the proposed algorithms in Sections 8.2.2 and 8.2.3. The information needed as input for the optimization algorithms are m, n, D_S_i, C_{ij}, T_{ij}, and DR_{ik} for $(1 \leq i \leq m, 1 \leq j \leq n)$. Those parameters are generated by a preprocessing program. with assumed certain distributions. Note here that n is the number of documents required by the user. For given m resources, D_S_i is distributed uniformly in $[0,1]$. Three different policies are applied to C_{ij}: free, a certain price per document, and varied price for each document (that is the case when the documents are paid by their sizes). For T_{ij}, every T_{i1} is bigger than T_{ij} $(j \geq 1)$, and $T_{i2} = T_{i3} = ... = T_{ij}$. Two different situations are considered for DR_{ik}. One is that all DR_{ik} for different i and k are distributed quite evenly in a certain range. The other is that resources are divided into several groups, with higher duplicates ratio for resources in the same group, and lower duplicates ratio for intergroup resources.

For a given set of inputs, we run those optimization algorithms and each of them will produce a solution $X(x_1, x_2,...,x_m)$ and $\sum_{i=1}^{m} x_i = n$, which includes a subset of resources and a number of documents that needs to be retrieved by each resource. The effectiveness of a solution could be evaluated by its utility value in the basic model with DR_{res} calculated by Equation 8.2.

Though the effectiveness of a solution could be evaluated by its utility value, it is still difficult to decide how good a solution is, since the utility value of a solution may vary considerably from one case to another. What we can do is to run different optimization algorithms based on the same input, then to make a comparison between them. Algorithm 8.1 can always generate the best solution and the worst solution (with a little change). It means that we can use it to obtain both the upper and the lower limits, but only for the situation of small scale experiment settings due to its time complexity. Another alternative for the lower limit is generated by a random method. In such a way, we can make the comparison more understandable by an indicator of normalized values in the range of $[0,1]$.

The first experiment looked at the effectiveness of each algorithm. The best and worst solutions were produced by Algorithm 8.1. Because Algorithm 8.1 is very time consuming, only 8-10 resources and 15-20 required documents were considered. Table 8.2 shows the average performance of every algorithm after 1000 runs.

In Table 8.2, Performance(1) is based on best-worst figures, that is, for method A, its performance is calculated by $P(1)_A = (s_A\text{-}s_{worst})/(s_{best} - s_{worst})$. Here s_A, s_{worst}, and s_{best} denotes the utility that method A obtains, worst utility score, and best utility score, respectively. Similarly, Performance(2) is based on best-random figures. Performance(2) of method A is calculated by $P(2)_A = (s_A - s_{random})/(s_{best} - s_{random})$. For the random method, we use the average utility scores of 100 random solutions for each run.

Table 8.2 Effectiveness of different models. Best serves as ceiling (100%) in both cases. In Performance(1), the worst solution serves as floor (0%); In Performance(2), 100 random solutions server as floor (0%).

Method	Performance(1)	Performance(2)
Static	0.948	0.819
Greedy	0.856	0.556
Dynamic	0.860	0.566
Random	0.686	—

In every run, the static model always outperforms both the greedy and dynamic methods, but it demands more time. The dynamic model is better than the greedy method in most runs. We distinguish between two typical situations. In the first, all the resource pairs have relatively even duplicates ratio, while in the second, resources are divided into several groups, with higher duplicates ratios for resource pairs in the same group and lower duplicates ratio for intergroup ones. The experiment suggests that in the first case both the greedy and the dynamic methods have very close performance, while in the second case the dynamic algorithm is better than the greedy algorithm considerably.

The second experiment aimed at comparing the performances of every algorithm with relatively larger number of resources (50, 100) and large number of documents required (300-400). Unfortunately, we can not afford the time for producing the best and worst solutions by running Algorithm 8.1. Instead, we use the random and static methods to serve as baselines. Although by doing so we do not know exactly the effectiveness of each method, we can however make a comparison of the four methods involved (static, greedy, dynamic, and random). The average results of 200 runs are shown in Table 8.3 (effectiveness comparison) and Table 8.4 (time needed).

In all the cases, the static model is the best and the dynamic model is better than the greedy method. On average, both the dynamic model and the greedy algorithm are 30% - 40% as effective as the static model. However, both the dynamic model and the greedy algorithm use only about 1% of the time that the static model needs.

Table 8.3 Effectiveness comparison of various models and the static model serves as ceiling (100%) and Random as floor (0%); in (x, y): x denotes the number of resources (50 or 100) and y denotes the number of documents (300, 350, or 400) required

Method	(50,300)	(50,350)	(50,400)	(100,300)	(100,350)	(100,400)
Greedy	29.8%	31.6%	33.2%	32.1%	31.3%	33.1%
Dynamic	36.9%	35.5%	38.1%	39.6%	37.1%	39.2%

Table 8.4 Time used by each model (in milliseconds); in (x, y): x denotes the number of resources (50 or 100) and y denotes the number of documents (300, 350, or 400) required

Method	(50,300)	(50,350)	(50,400)	(100,300)	(100,350)	(100,400)
Static	161	205	299	1306	1342	1484
Greedy	1	2	3	8	9	11
Dynamic	3	3	4	9	12	14

8.2.5 Further Discussion

In our discussion about the multi-objective model for resource selection in Sections 8.2.2-8.2.4, four objectives are considered. They are relevance, monetary cost, time, and duplicates ratio. Among these four objectives, relevance and duplicates ratio have very close relation, while the two others are independent. If considering relevance and duplicates ratio together, we may be able to find more effective solutions for the resource selection problem. Let us look at an example to illuster this.

Example 8.2.1. There are four overlapping databases D_a, D_b, D_c, and D_d. For a given query q, there are 7, 6, 5, and 3 relevant documents in D_a, D_b, D_c, and D_d, respectively. But there are some overlaps. See the graph below.

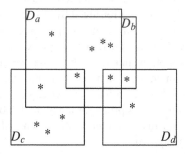

In the above graph, each database is represented by a square, and each relevant document is represented by a star. If a star is located inside a square, then it means that the relevant document is stored in that database. Some relevant documents exist in more than one database. Suppose that the user would obtain as many relevant documents as possible by accessing 2 databases, if we do not consider the duplicate problem then selecting both databases D_a and D_b is the best option ($7 + 6 = 13$ in total). The second best is to select D_a and D_c, we can obtain 12, and the next one is D_b and D_c, we can obtain 11. However, if duplicates are not counted, we can actually obtain 8 relevant documents from D_a and D_b; we can obtian 10 relevant documents from D_a and D_c, or D_b and D_c. Both options (D_a and D_c, or D_b and D_c) can provide more relevant documents for us than the first option (D_a and D_b).   □

From the above example, we can see that resource selection in the condition of overlapping databases becomes a maximum covering problem. It is a classical question in computer science because it is NP-hard. Interested readers may refer to [29, 47, 22] for more information.

8.3 Results Merging

For overlapping databases, especially when the overlap is considerable, it is not good for us to use those result merging methods that are proposed for pair-wise disjoint databases. In the case of overlapping databases, duplicate documents occuring in multiple databases are useful for at least two different aspects. Firstly, they are helpful for correlating the scores of all the documents contained in different results to make scores across different search engines comparable. For example, as in [18, 85], regression analysis can be used for achieving this. The same can be done as long as there are overlaps among databases, no matter what the overlap rate is. Secondly, we can still use "the multiple evidence principle" as in data fusion to improve the performance of result merging. However, two further points should be noticed. Firstly, "the multiple evidence principle" can only be effectively applied when the overlap among databases is considerable; secondly, we may need different result merging methods for overlapping databases from the ones used for data fusion. Let us illustrate this by an example. Suppose one document d_1 is retrieved from database A, and another document d_2 is retrieved from both databases A and B, if we know that A and B are identical, which is the situation for data fusion, then we are sure that d_1 is stored in B but not retrieved by B's search engine. If A and B are not identical, there are two possibilities: either d_1 is stored in B but not retrieved, or d_1 is not stored in B at all. However, we do not know exactly in which situation it is.

Based on different assumptions that could be made, we may have different kinds of solutions. For any document d, which occurs in database A's result but not B's result, if we always assume that d is in B but not retrieved by B's server (Assumption 1), then such an assumption is appropriate for identical databases or nearly identical databases. Actually, all data fusion methods with identical databases rely on this since the assumption is always true.

For overlapping databases, especially when the overlap is not very heavy, the above assumption is error-prone. We may take another assumption (Assumption 2). That is, if d is not retrieved by B's server, we simply assume it is not stored in B.

The third option is dynamic estimation. We may take one of the two above assumptions depending on certain circumstance. For a better estimation, the overlap rate between two databases is a piece of useful information. However, usually it is quite difficult and very costly to obtain accurate information about that in a federated

search environment, since it is possible that all component databases are always under change, and on the other hand it is not common that database servers are willing
to provide interfaces to the public for accessing their databases directly. Therefore,
for the discussion in this section, we assume that no extra information is available
except the results themselves.

All data fusion methods can be regarded as taking Assumption 1, while all result
merging methods investigated in this section take Assumption 2. As experiments
in [126, 116, 124] suggests: Assumption 2 is more suitable than Assumption 1 for
overlapping databases, even when the overlap is quite heavy. The reason for that is
based on the following hypothesis:

For merging results from overlapping databases, when we take Assumption 1,
the benefit we obtain from a correct estimation is less than the damage we obtain
from a wrong estimation.

This hypothesis can explain why some data fusion methods such as CombMNZ
deteriorate very quickly when the databases deviate from identical gradually. As
we shall see later in this section, CombMNZ performs quite poorly even when the
overlap is very heavy (e.g., as defined in Section 8.3.2, overlap rate (see Equation
8.10) among databases is up to 80%).

When we take Assumption 2, one thing we can do for a more accurate estimation
is to retrieve more documents in every component database. For example, suppose
the federated search system needs to return 100 documents to the user for a given
query, and 4 component databases are available, then usually retrieving 30 to 40
documents from each database is enough. However, if we retrieve more documents
from each database, then we know with more certainty that those top ranked documents from each database are stored in other databases or not. In Section 8.3.3 we
shall discuss this in more detail.

Another problem we are facing is how to merge documents that are retrieved by
different number of databases. It is a new situation that does not happen in data
fusion with identical databases. For example, if d_1 is only stored in A, d_2 is stored
in both B and C; for a given query, d_1 is retrieved by A with a score of 0.6, d_2 is
retrieved by B with a score of 0.4 and C with a score of 0.4. How to rank these two
documents is a key issue.

8.3.1   Several Result Merging Methods

In the following, we introduce two results merging methods that can be used for
overlapping databases. They are the shadow document method (SDM) and the multi-
evidence method (MEM). SDM and MEM assume that every document retrieved
from any available database has a score which illustrates the estimated relevance of
that document to the given query.

a. The Shadow Document Method (SDM)
The shadow document method (SDM) works like this. For a given query q and two
component databases A and B, if document d occurs in both A and B's results then

we sum these two scores as d's global score; if d only occurs in A's result with a score s_1, then it is not stored in B (according to Assumption 2); and we imagine if d is stored in B, it is very likely to be retrieved with a score close to s_1 (normalized). For the shadow document of d in B, we assign a score to that (shadow) document of d in B. We do the same for those documents that only occurs in B's result but not in A's result. Then we are able to use CombSum to merge all the results. More generally, if we have a query q and n databases D_i $(1 \leq i \leq n)$ for retrieving, and document d occurs in m databases with score s_i $(1 \leq i \leq m \leq n)$ respectively, then d's total score is:

$$score(d) = \sum_{i=1}^{m} s_i(d) + k_1 \frac{n-m}{m} \sum_{i=1}^{m} s_i(d) \tag{8.9}$$

where k is a coefficient $(0 \leq k_1 \leq 1)$. For each result without d, we assign it a value that is a certain percentage (k_1) of the average score of d from m databases. To assign a desirable value to k_1 is important and it can be determined empirically.

Some modifications can be done to improve the above-mentioned method. One way is to let the coefficient k_1 be related to the overlapping rate of two databases. We need some statistics about these databases such as their sizes, overlapping rates [118] and so on. However, to collect such information usually takes a lot of effort. If such statistics are not available, we may still be able to collect some useful information on the fly. When merging results for a given query, we calculate the number of duplicate documents between each pair of them, and then use this piece of information as the indicator of the overlapping rate of the two databases. Let us consider the situation of only two databases D_1 and D_2. Suppose both results comprise equal number m_1 of documents and the number of duplicates between them is m_2. We can rely on this to modify the above-mentioned method. If document d only occurs in D_1 and obtain a score of d_1, then the total score of d is calculated as $s_1(1 + k_2 * (m_1 - m_2)/m_1)$, where k_2 is a coefficient $(0 \leq k_2 \leq 1)$. The same as for k_1, we need to find a suitable value for k_2 empirically by carrying out some experiments. It is straightforward to modify the algorithm for more databases.

If we clearly know which document is stored in which database, then we may be able to work out a more effective merging method for those results involved. This might be possible in some situations, for example, corporate networks. Let us discuss this with two databases D_1 and D_2 again. For a given query q, assume that a document d only occurs in the result L_1 from D_1 with a score of s_1, but not L_2 from D_2. If d is not stored in D_2, then the score of d is $2s_1$; If d is stored in D_2 but not retrieved, then the total score of d is just s_1.

Furthermore, we can estimate the performance of each database by some sample queries or based on previous retrieval experience, then each database obtains a score corresponding to its estimated performance. The shadow document method can also be improved by this database score as in the linear combination data fusion method. However, we don't discuss weighted result merging methods further in this section, since most methods discussed can be expanded with weighting by some methods or variants of them discussed in Chapter 5.

b. The Multi-Evidence Method (MEM)

In this method, we first average the score of every document, then multiply that score by a factor $f(i)$, which is a function of the number of databases that include the document in their results. The value of $f(i)$ increases with i, which indicates the increasing evidence about the relevancy of the document to the given query when more search engines have a common opinion. For example, if d_1 is retrieved by A with a (normalized) score of 0.6, and retrieved by B with a (normalized) score of 0.5, then its combined score is $f(2) * (0.6 + 0.5)/2$. How to determine $f(i)$ for $(i = 1, 2, 3, ...)$ is important to this method.

If we use different $f(i)$, we may obtain very different result merging methods. For example, if we let $f(i) = 1$ for $(i = 1, 2, 3, ...)$, then the multi-evidence method is just equal to the averaging method. If we let $f(i) = i$ for $(i = 1, 2, 3, ...)$, then it is CombSum; if we let $f(i) = i^2$ for $(i = 1, 2, 3, ...)$, then it is CombMNZ.

Lee [48] did an experiment for data fusion with six text search engines. He let $f(i) = avg\_score * i^\beta$ with different β values (1, 1.5, 2, 3, 6, and 11) assigned to $f(i)$. He found that it worked best when $\beta = 2$, which was exactly the CombMNZ method. It suggests that with identical databases, we should set a weight as heavy as i^2 for those documents retrieved by multiple resources. However, for overlapping databases, the situation is different. $f(i)$ should be defined with a value between 0 and 2 for its β.

A more sophisticated solution is to let $f(i)$ vary with the degree of overlap among databases. If the overlap is heavy, we may use a large β; if the overlap is light, we may use a small β. However, we only consider static $f(i)$ which could be used in all different situations. We leave the adaptable $f(i)$ issue for further investigation.

Note that both SEM and MEM work with scores, which need to be normalized in many cases. In Chapter 3, various types of score normalization methods were discussed. Generally speaking, any of the methods disussed there can also be used for this purpose.

8.3.2 Evaluation of Result Merging Methods

Here we assume that no extra information is available except the resultant lists from different resources. For example, we do not know if any particular document is stored in any particular database, and we also do not know any statistics about the sizes of databases and overlapping rate of these databases, and so on. Besides, we assume that every search engine performs equally well.

In the experiment, both scores and rankings of documents in resultant lists are considered. Two differnt score normalization methods are used: the zero-one linear normalization and the linear regression normalization. We hope to observe the difference caused by these two different normalization methods to result merging.

TREC text collections and topics were used. 50 TREC topics (251-300) and 524,929 documents (AP88, CR93, FR88, FR94, FT91-94, WSJ90-92, and ZF, which comprised the collection used by the TREC 5 ad hoc track [101]).

Both Lemur [49] and Lucene [54] information retrieval tool kits were used. Lemur provides three options as retrieval models: vector space, okapi, and the language model, while Lucene provides two options, vector space and the probabilistic model[1], as retrieval models.

All the documents were divided into 250 partitions. Each included around 2,000 documents. The generation of 5 databases is as follows: each partition is included in at least one database, then it is guaranteed that the number of different documents in all 5 databases is invariable. We used a random process to determine which database(s) to include which partition. However, an adjustable parameter was used to help control the overlapping rate among 5 databases. After generation, they were indexed and retrieved with the same TREC queries. Each database is working with a different retrieval model in either Lemur or Lucene. We consider this is more likely to be the case for real applications, especially in loosely coupled federated search systems. Then their results were merged using different merging methods.

In the experiment, 4 methods including Round-robin, CombMNZ, SDM (the basic one), and MEM, were tested. Round-robin served as the baseline. Round-robin works in a style that it takes documents in trun from all available databases. For MEM, the zero-one score normalization method was used. For SDM, three different types of score normalization methods, which were the zero-one method, Borda count, and Bayesian inference-based method, were used. They are referred to as SDM, the Borda count [3], and the Bayesian fusion [3], respectively. Thus, it seems that we have 6 methods in total. In all these methods, both Round-robin and the Borda count use rankings, while CombMNZ, SDM, and MEM use scores. the Bayesian fusion may use either rankings or scores. CombMNZ is the method for data fusion, it is tested here in order to make a comparison between this method and those for overlapping databases. Despite having the same names, Borda count and Bayesian fusion are result merging methods for overlapping databases. They are not the same as Borda count and Bayesian fusion used for data fusion.

As indicated in Section 8.3.1, some of the methods need to set a few parameters. For the Bayesian fusion, we first normalize scores of documents into the range of [0, 1], then convert these normalized scores into possibilities [0, 0.5]. Here 0.5 is chosen arbitrarily without any knowledge of the search engines used. In such a way, all the scores calculated in the Bayesian fusion are negative, parameter k needs to be set no less than 1. 21 different values (1, 1.05, 1.1, ..., 3) were tested in the experiment. For both Borda count and SDM, $k = 0.0, 0.05, 0.1, ..., 1$ were tested. For MEM, we let $f(i) = 1 + log(i)$.

For a fair comparison of the performance of all result merging methods, it is necessary to set these parameters in a way that does not favour any particular method. For each parameter, we choose a typical value for it. $k = 0.5$ is chosen for SEM, $k = 0.95$ for the Borda count, and $k = 2.5$ for the Bayesian fusion. These values usually enable corresponding result merging methods to produce good but not the best results. Besides, the effect of k on the performance of two methods (SDM and

---

[1] The probabilistic model was implemented by the author.

Bayesian fusion) is analyzed in detail later in this section. We shall see that these choices for parameters setting are reasonable.

We use o_rate to measure the degree of overlap among a group of databases. It is defined as follows:

$$o\_rate = \frac{(\sum_{i=1}^n |D_i|) - |D_{all}|}{(n-1) * |D_{all}|} \tag{8.10}$$

where n is the number of databases involved, $|D_i|$ is the number of documents in database D_i, and $|D_{all}|$ is the total number of different documents in all n databases. The definition of o_rate in Equation 8.10 is a little different from the one in Section 4 (see Equation 4.1).

When there is no duplicates, $\sum_i^n |D_i| = |D_{all}|$, $o\_rate = 0$; when all the databases have the same documents, $|D_i| = |D_{all}|$ for ($i = 1, 2, ..., n$), then $o\_rate = 100\%$.

For example, if we have 100,000 different documents in total, database D_1 comprises 70,000 documents, D_2 comprises 50,000 documents, and D_3 comprise 30,000 documents, then $o\_rate = (70000+50000+30000-100000)/(2*100000) = 25\%$.

We used a different model in either Lemur or Lucene to retrieve each database in a quintuple. Different result merging methods were evaluated by using Round-robin as baseline. We divide the range [0,1] of o_rate into five equal intervals [0, 0.2], (0.2, 0.4], (0.4, 0.6]. (0.6, 0.8], and (0.8, 1]. In each inverval, 10 runs were tested. In each run, five databases with certain overlapping were generated. Each time the overlap rate is a randomly selected number in a given interval.

Some of the experimental results are shown in Tables 8.5-8.9. The zero-one linear score normalization is used for SDM, MEM, and CombMNZ. Because all five retrieval models in Lemur and Lucene used in the experiment are quite good and close in performance, the relevant documents are evenly distributed in each of the five databases, and the total number of documents in all 5 quintuples are always the same, the performance of Round-robin is very stable no matter what the overlap rate is among these databases. Compared with Round-robin, all other methods do better in the top 5 or 10 ranked documents than more documents. CombMNZ deteriorates more quickly and deeply than any other method in performance when the overlap rate falls. It is not surprising since all these methods exploit the multiple evidence principle to boost the rankings of those commonly retrieved documents, while Round-robin does not. Because CombMNZ is the heaviest to use this principle, it suffers the most when the overlap rate is low. In the same situation, SDM, MEM, Borda fusion, and Bayesian fusion do not suffer so much since a balanced mechanism is introduced.

In the following let us discuss the experimental result in two groups, each with a different normalization method. First with the zero-one normalization, Figure 8.6 shows the average precision of 6 methods at 8 document levels (5, 10, 15, 20, 25, 30, 50, and 100) with different overlap rates. The linear zero-one normalization method is used. Both SDM and MEM are close in performance in most situations. They perform better than Round-robin when the overlap rate among databases is no less than 40% (though the differences are not significant sometimes), they are close to Round-robin when the overlap rate is between 20% and 40%, and they become not as good as Round-robin when the overlap rate is less than 20% (the differences are

Table 8.5 Average precision of different methods with overlap rate between 0%-20% (the zero-one linear normalization for SDM, MEM, and CombMNZ, Round-robin serves as baseline)

Rank	Round-robin	CombMNZ	Bayesian (k=2.5)	Borda (k=0.95)	SDM (k=0.5)	MEM
5	**0.2676**	0.2436*	0.2336*	0.2448*	0.2628	0.2648
		(-9.0%)	(-12.7%)	(-8.5%)	(-1.8%)	(-1.0%)
10	**0.2437**	0.2050*	0.2232*	0.2280*	0.2362	0.2436
		(-15.9%)	(-8.4%)	(-6.4%)	(-3.1%)	($\pm$ 0.0%)
15	**0.2228**	0.1860*	0.2081*	0.2131*	0.2173	0.2197
		(-16.5%)	(-6.6%)	(-4.4%)	(-2.5%)	(-1.4%)
20	**0.2079**	0.1681*	0.1962*	0.1987*	0.2021	0.2023
		(-19.1%)	(-5.6%)	(-4.4%)	(-2.8%)	(-2.7%)
25	**0.1953**	0.1573*	0.1852*	0.1852*	0.1883	0.1882*
		(-19.5%)	(-5.2%)	(-5.2%)	(-3.6%)	(-3.6%)
30	**0.1830**	0.1464*	0.1724*	0.1764*	0.1789	0.1798
		(-20.0%)	(-5.8%)	(-3.6%)	(-2.2%)	(-1.7%)
50	**0.1542**	0.1198*	0.1466*	0.1488*	0.1520	0.1516*
		(-22.3%)	(-4.9%)	(-3.5%)	(-1.4%)	(-1.7%)
100	**0.1181**	0.0900*	0.1123*	0.1127*	0.1129*	0.1136*
		(-23.8%)	(-4.9%)	(-4.6%)	(-4.4%)	(-3.8%)

Table 8.6 Average precision of different methods with overlap rate between 20%-40% (the zero-one linear normalization for SDM, MEM, and CombMNZ, Round-robin serves as baseline)

Rank	Round-robin	CombMNZ	Bayesian (k=2.5)	Borda (k=0.95)	SDM (k=0.5)	MEM
5	0.2844	0.2324*	0.2614*	0.2502*	0.2862	**0.2916**
		(-18.3%)	(-21.9%)	(-8.1%)	(+0.6%)	(+2.5%)
10	0.2584	0.2004*	0.2392*	0.2372*	0.2558	**0.2587**
		(-22.4%)	(-7.4%)	(-8.2%)	(-1.0%)	(+0.1%)
15	0.2352	0.1780*	0.2162*	0.2187*	0.2338	**0.2362**
		(-24.3%)	(-8.1%)	(-7.1%)	(-0.6%)	(+0.4%)
20	0.2169	0.1632*	0.2017*	0.2074*	0.2158	**0.2194**
		(-24.8%)	(-7.0%)	(-4.3%)	(-0.5%)	(+0.1%)
25	0.2015	0.1519*	0.1897*	0.1950*	0.2012	**0.2026**
		(-24.6%)	(-5.6%)	(-3.2%)	(-0.1%)	(+0.5%)
30	0.1899	0.1519*	0.1788*	0.1833*	0.1888	**0.1917**
		(-20.0%)	(-5.8%)	(-3.5%)	(-0.6%)	(+0.9%)
50	0.1596	0.1438*	0.1505*	0.1525*	0.1588	**0.1597**
		(-9.9%)	(-5.7%)	(-4.4%)	(-0.5%)	($\pm$0.0%)
100	**0.1224**	0.0948*	0.1138*	0.1159*	0.1183*	0.1203
		(-22.5%)	(-7.0%)	(-5.3%)	(-3.3%)	(-1.7%)

Table 8.7 Average precision of different methods with overlap rate between 40%-60% (the zero-one linear normalization for SDM, MEM, and CombMNZ, Round-robin serves as baseline)

Rank	Round -robin	CombMNZ	Bayesian (k=2.5)	Borda (k=0.95)	SDM (k=0.5)	MEM
5	0.2782	0.2812	0.2950*	0.2794	0.3157*	**0.3163***
		(+1.6%)	(+6.1%)	(+0.6%)	(+13.5%)	(+13.7%)
10	0.2612	0.2411*	0.2618	0.2495*	0.2746*	**0.2772***
		(-7.7%)	(+0.2%)	(-4.5%)	(+5.1%)	(+6.2%)
15	0.2376	0.2179*	0.2385	0.2279*	0.2505*	**0.2541***
		(-8.3%)	(-0.4%)	(-4.1%)	(+5.4%)	(+6.9%)
20	0.2250	0.1995*	0.2189	0.2085*	0.2328*	**0.2335***
		(-11.3%)	(-2.7%)	(-7.3%)	(+3.3%)	(+3.8%)
25	0.2130	0.1860*	0.2039*	0.1940*	0.2161	**0.2182**
		(-12.7%)	(-4.3%)	(-8.9%)	(+1.5%)	(+2.4%)
30	0.1967	0.1750*	0.1901	0.1836*	0.2047*	**0.2063***
		(-11.0%)	(-3.4%)	(-6.7%)	(+4.1%)	(+4.9%)
50	0.1624	0.1489*	0.1604	0.1546*	0.1726*	**0.1739***
		(-8.3%)	(-1.2%)	(-4.8%)	(+6.3%)	(+7.1%)
100	0.1264	0.1143*	0.1213*	0.1153*	0.1277	**0.1293**
		(-9.6%)	(-4.0%)	(-8.8%)	(+1.0%)	(+2.3%)

Table 8.8 Average precision of different methods with overlap rate between 60%-80% (the zero-one linear normalization for SDM, MEM, and CombMNZ, Round-robin serves as baseline)

Rank	Round -robin	CombMNZ	Bayesian (k=2.5)	Borda (k=0.95)	SDM (k=0.5)	MEM
5	0.2760	0.2939*	0.3013*	0.2819	**0.3139***	0.3083*
		(+6.5%)	(+9.2%)	(+2.1%)	(+13.7%)	(+11.7%)
10	0.2571	0.2516	0.2569	0.2445*	0.2740*	**0.2759***
		(-2.1%)	(± 0.0%)	(-4.9%)	(+6.6%)	(+7.3%)
15	0.2363	0.2212*	0.2365	0.2223*	0.2488*	**0.2500***
		(-6.4%)	(±0.0%)	(-5.9%)	(+5.3%)	(+5.8%)
20	0.2223	0.2009*	0.2163	0.2086*	0.2316*	**0.2334***
		(-9.6%)	(-2.7%)	(-6.1%)	(+4.2%)	(+4.9%)
25	0.2081	0.1877*	0.2048*	0.1936*	**0.2172***	**0.2172***
		(-9.8%)	(-1.6%)	(-7.0%)	(+4.4%)	(+4.4%)
30	0.1948	0.1778*	0.1880*	0.1823*	0.2035*	**0.2059***
		(-8.7%)	(-3.5%)	(-6.4%)	(+4.7%)	(+5.6%)
50	0.1610	0.1494*	0.1597	0.1548*	0.1720*	**0.1736***
		(-7.2%)	(-0.8%)	(-3.9%)	(+6.8%)	(+7.8%)
100	0.1259	0.1115*	0.1180*	0.1149*	0.1275	**0.1290***
		(-11.4%)	(-6.3%)	(-8.7%)	(+1.3%)	(+2.5%)

Table 8.9 Average precision of different methods with overlap rate between 80%-100% (the zero-one linear normalization for SDM, MEM, and CombMNZ, Round-robin serves as baseline)

Rank	Round-robin	CombMNZ	Bayesian (k=2.5)	Borda (k=0.95)	SDM (k=0.5)	MEM
5	0.2738	0.3032*	0.3156*	0.2956*	0.3120*	**0.3188***
		(+10.7%)	(+15.3%)	(+8.0%)	(+14.0%)	(+16.4%)
10	0.2600	0.2691	0.2678	0.2482*	**0.2753***	0.2726*
		(+3.5%)	(+3.0%)	(-4.5%)	(+5.9%)	(+4.8%)
15	0.2390	0.2399	0.2411	0.2224*	**0.2458***	0.2443
		(+0.4%)	(+0.9%)	(-6.9%)	(+2.8%)	(+2.2%)
20	0.2258	0.2174	0.2234	0.2064*	0.2292	**0.2296**
		(-3.7%)	(-1.1%)	(-8.6%)	(+1.5%)	(+1.7%)
25	0.2129	0.2032*	0.2082	0.1908*	0.2149	**0.2163**
		(-4.6%)	(-2.2%)	(-10.4%)	(+0.9%)	(+1.6%)
30	0.1991	0.1893*	0.1946	0.1805*	**0.2044**	0.2039
		(-4.9%)	(-2.3%)	(-9.3%)	(+2.7%)	(+2.4%)
50	0.1635	0.1612	0.1656	0.1521*	**0.1731***	0.1724*
		(-1.4%)	(+1.3%)	(-7.0%)	(+5.9%)	(+5.4%)
100	0.1263	0.1214	0.1238	0.1124*	**0.1300**	0.1298
		(-3.9%)	(-2.0%)	(-11.0%)	(+2.9%)	(+2.8%)

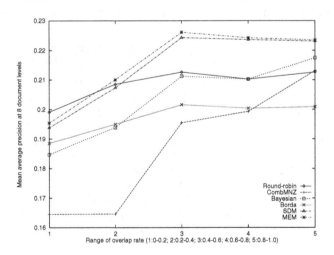

Fig. 8.6 Performances of 6 methods with different overlap rates (zero-one normalization)

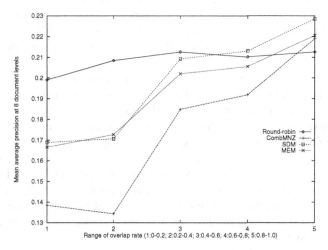

Fig. 8.7 Performances of 4 methods with different overlap rates (regression normalization)

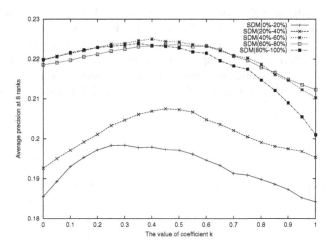

Fig. 8.8 SDM's average performance with different coefficient values

not significant in most cases). CombMNZ is not as good as SDM and MEM in all cases (the maximal overlap rate used is about 95% in the experiment). However, we do not try to find the necessary conditions for CombMNZ to be better than SDM and MEM. If it exists, it must be very close to 100%. Bayesian fusion does not work as well as SDM and MEM. Its performance is slightly better than Round-robin when the overlap rate is more than 80%. Borda fusion is not as good as Round-robin in all situations.

Figure 8.7 shows the average precision of 4 methods at 8 document levels (5, 10, 15, 20, 25, 30, 50, and 100) with linear regression score normalization. SDM and

MEM perform slightly better than Round-robin when overlap rate is above 80%. When overlap rate is between 60% and 80%, the performances of SDM and Round-robin are close. CombMNZ's performance is close to Round-robin when overlap rate is over 80%. In all other situations, Round-robin outperforms all three other methods.

In summary, both SDM and MEM perform better with the zero-one linear normalization than with linear regression normalization in most situations and on average. The only exception we observe is SDM with overlap rates between 80% and 100%. Moreover, using zero-one linear normalization, all result merging methods are more stable in performance under different overlap rates than using linear regression normalization. Actually, such a phenomenon is quite surprising, since in [116], linear regression normalization was successfully used with the SDM method. The only observable difference is that only Lemur was used in [116], while both Lemur and Lucene have been used here. It suggests that the zero-one linear normalization is a better option than linear regression normalization for result merging in heterogeneous environment.

Figure 8.8 shows the average precision at 8 document levels (5, 10, 15, 20, 25, 30, 50, and 100) of SDM (zero-one linear normalization) with different overlap rate and different coefficient values of k. We can observe that SDM's performance is the lowest when overlap rate is between 0% and 20%, then it is better when overlap rate increases to between 20% and 40%, and it is the best when overlap rate is above 40%. However, when overlap rate is no less than 40%, further increase in overlap rate is not contributive for increasing retrieval performance. Compare the three curves for SDM(40%-60%), SDM(60%-80%), and SDM(80%-100%), they

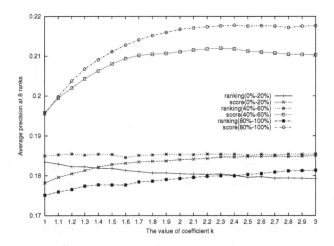

Fig. 8.9 Comparison of using score and ranking in Bayesian fusion

are very close to each other or even winding together. We observe that the same phenomenon happens to MEM, Borda fusion, and Bayesian fusion with different overlap rates (see also Figure 8.6).

In addition, Figure 8.8 shows the effect of different k on the performance of SDM. It suggests that a value between 0.2 and 0.6 for k is acceptable.

We experimented with Bayesian fusion using score and ranking respectively (figures shown in Table 8.5-8.9 using scores). Figure 8.9 shows the average precision of Bayesian fusion with three different overlap rates (0%-20%, 40%-60%, 80%-100%). With ranking information only, all three curves are very close; while significant difference can be observed for three curves using score information. The "score(0%-20%)" one is the lowest, the "score(40%-60%)" one is in the middle, while the "score(80%-100%)" one is the highest. Let us take a look at them in three pairs, each with the same overlap rate but different available information (score or ranking). For the pair with overlap rate between 0%-20%, the curves are quite close to each other. For the two other pairs (overlap rate between 40%-60% and between 80%-100%), the difference between the two curves is quite large (10%-15%). Since all other aspects are completely the same, it demonstrates that score is more informative than ranking in result merging.

8.3.3   Overlap between Different Results

We hypothesize that all "good" text retrieval servers are very similar from a user's point of view. That is to say, all "good" text retrieval servers are prone to retrieve the same documents for the same query. Generally speaking, the number of duplicates in two different results may depends on several aspects:

1. The number of documents in each result;
2. The size and content of each database;
3. The size of overlapping part in two databases;
4. search engines used in two databases.

In this subsection, we do not intend to study the effect of all these aspects to the duplicated documents in different results. More specifically, we consider it with overlapping databases. One experiment was conducted like this:

1. Creating two overlapping databases, in which C is the common document set;
2. Using two different retrieval models and the same query to retrieve both databases so as to get two different results L_a and L_b;
3. Picking up those documents in the top n of L_a which belong to C, and check how many of them appear in the top n, $2n$, $3n$,... documents of L_b;
4. Changing the positions of L_a and L_b and do the same thing as in the last step.

Figure 8.10 shows the relationship of overlap between two retrieved results, which is based on the average of 50 queries over a pair of overlapping databases. Two different retrieval models of Lemur were used for the retrieval of two databases.

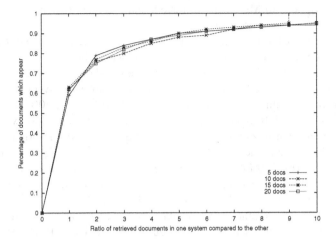

Fig. 8.10 Overlap between two results for the same query

We took the top n (n=5, 10, 15, and 20) documents in one result, picked up those documents from the overlapping part, then check how many of them appear in the top n, $2n$, ..., $10n$ in the other result, the percentages of appearing are drawn in the plot, in which four curves overlap considerably.

For documents that come from the overlapping part and locate in the top n(n= 5, 10, 15, or 20) of one result, about 60% of them appear in the top n and over 90% of them appear in the top $10n$ of the other result. The curves in both directions are almost the same. Furthermore, 12 more pairs of overlapping databases with different overlap rates and different documents were tested, we obtained very similar results as in Figure 8.10 for all of them. When we used the same model of Lemur to retrieve these database pairs, then the percentages were higher than that shown in Figure 8.10.

From this experiment, we observe that if we require n documents for the merged results, then to retrieve more documents (for example, $8n$ to $10n$ documents) is a good policy that enables us to estimate more accurately if a document exists in a particular database or not. Therefore, retrieving more documents from each component database can improve the performance of result merging methods accordingly.

8.3.4 Conclusive Remarks

In Section 8.3.2 we have experimented with several result merging methods in federated search environments. TREC document collections have been used and diversified overlaps among databases have been generated to evaluate the performance of these result merging methods on average precision at different document levels. SDM, MEM, Bayesian fusion, Borda fusion, CombMNZ, and Round-robin (served as baseline), have been evaluated. Two score normalization methods, which are the zero-one linear normalization and linear regression normalization, and two different kinds of available information, which are scores and rankings, have been tested. From these experiments, we have several observations as follows:

1. Experimental results show that CombMNZ, SDM and MEM perform better with the zero-one linear normalization than with the linear regression normalization;
2. If we only consider the top 10 documents, then all five methods (CombMNZ, Borda fusion, Bayesian fusion, SDM, and MEM) perform better than Round-robin when overlap rate is above 60%; for SDM and MEM, the above requirement can be reduced to 40%. Even when the overlap rate is below 40%, the performances of both SDM and MEM are very close to that of Round-robin.
3. By comparing the performance of Bayesian fusion with two kinds of information, score and ranking, we demonstrate that score is more informative than ranking for result merging;
4. The experiment also shows that both SDM and MEM are effective methods in most situations. They perform better than Round-robin when the overlap rate is no less than 40%.
5. Borda fusion is not as good as Round-robin in all cases;
6. Both Bayesian fusion and CombMNZ are not as good as Round-robin when overlap rate is below 80%. When overlap rate is above 80%, CombMNZ and Round-robin become comparable, while Bayesian fusion is slightly better than Round-robin;
7. Round-robin is quite good in all situations since the retrieval servers used in the experiment are quite close in performance and relevant documents are evenly distributed in every database.

In order to achieve better performance for result merging methods such as the shadow document method, the multi-evidence method, and Bayesian fusion, more documents than that appear in the merged result are required from each component database. For example, if 20 documents are needed in the final result, then retrieving about 200 documents from each component database may lead to more effective merging than using fewer documents for them.

Some methods such as CombMNZ that are very good for data fusion with identical databases are very poor for result merging with overlapping databases. Even when the overlap is quite heavy (e.g., when the overlap rate among databases is up to 60%), CombMNZ's performance is not good at all. On the other hand, methods such as CombMNZ are not suitable for overlapping databases since they only pick up those documents stored in all or most of the databases but ignore those documents stored in only one or very few databases. Therefore, such merging methods are not suitable on the aspect of coverage.

As previous discussion in Chapters 4-7 demonstrates that the strength of correlation among component results affects data fusion considerably, we have good reason to hypothesize that is the case for result merging from overlapping databases. The experimental environment setting in Section 8.3.2 is modestly heterogeneous – two different retrieval systems, one with three different models, and the other with two different models. Two other situations, homogeneous and extremely heterogeneous environments, are also worth investigation. In addition, test with some other kinds of collections, for example, in a language other than English, or on a different topic other than news articles, maybe helpful for us to understand more about the result merging issue.

Chapter 9
Application of the Data Fusion Technique

Different aspects of the data fusion technique have been addressed so far. Surprisingly, data fusion can be used in many different situations. In this chapter, we are going to discuss some applications of it.

9.1 Ranking Information Retrieval Systems with Partial Relevance Judgment

Since the web and digital libraries have more and more documents on these days, there is a need to test and evaluate information retrieval systems with larger and larger collections. In such a situation, how to make the human judgment effort reasonably low becomes a major issue. Partial relevance judgment is a possible solution to this in information retrieval evaluation events such as TREC and NTCIR. However, some further questions come up:

- What is the effect of partial relevance judgment on the evaluation process and system performance?
- Are there any partial relevance judgment methods other than the pool strategy can be applied?
- Which measures should we use for such a process?
- What can be done to make the evaluation process more reliable in the condition of partial relevance judgment?

In order to answer some of these questions, some researchers [15, 77] tried to define some new measures that are suitable for partial relevance judgment. Besides the pooling strategy, some other partial relevance judgment methods were also investigated. Here we focus on how to rank a group of information retrieval systems based on the pooling method and partial relevance judgment, since the pooling method has been used in TREC and NTCIR and other information retrieval evaluation events for many years.

S. Wu: Data Fusion in Information Retrieval, ALO 13, pp. 181–212.
springerlink.com   © Springer-Verlag Berlin Heidelberg 2012

Table 9.1 Information summary of 9 groups of results submitted to TREC

Group	Track	Number of results	Number of topics
TREC 5	ad hoc	61	50
TREC 6	ad hoc	71*	50
TREC 7	ad hoc	103	50
TREC 8	ad hoc	129	50
TREC 9	Web	105	50
TREC 2001	Web	97	50
TREC 2002	Web	71	50
TREC 2003	robust	78	100
TREC 2004	robust	101	249**

Note: *Three results submitted to TREC 6 are removed since they include far fewer documents than the others.
**One topic in TREC 2004 is dropped since it does not include any relevant document.

9.1.1   How Is the Problem Raised?

First let us investigate the effect of partial relevance judgment on those commonly used system-oriented metrics such as average precision and recall-level precision. We carry out an empirical study with TREC data. Since almost in all the cases, TREC uses partial relevance judgment. Ideally, the experiment should compare the performances of the same information retrieval system with complete relevance judgment and with partial relevance judgment. If there is not much difference, then that is very good. We can comfortably use the result from partial relevance judgment as an estimate for the complete relevance judgment. Otherwise, we need to find what the difference is between them and figure out what can be expected with complete relevance judgment if only partial relevance judgment is available. However, such an experiment is no doubt very expensive because complete relevance judgment is required. We carry out an experiment in the opposite direction. This will be explained in detail below.

Considering that the pooling method in TREC is a reasonable method for partial relevance judgment, we conduct an experiment to compare the values of these metrics by using pools of different depths. In every year, a pool of 100 documents in depth was used in TREC to generate its *qrels* (relevance judgment file). Shallower pools of 10, 20,.., 90 documents in depth were used in this experiment to generate more *qrels*. For a resultant list and a metric, we calculate its value of the metric c_{100} using the 100 document *qrels*, then calculate its value of the measure c_i using the i document *qerls* ($i = 10, 20, ..., 90$), their absolute difference can be calculated using $abs\_diff = |c_i - c_{100}|$ and their relative difference can be calculated using $rel\_diff = |c_i - c_{100}|/(c_{100})$.

9 groups of runs submitted to TREC (TREC 5-8: ad hoc track; TREC 9, 2001, and 2002: web track; TREC 2003 and 2004: robust track) were used in the experiment. Their information is summarised in Table 9.1.

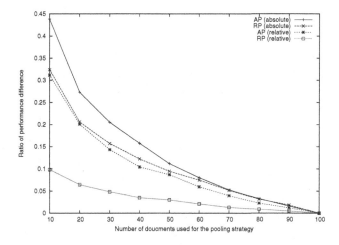

Fig. 9.1 Relative and absolute differences of two metrics when using pools of different depths (the pool of 100 documents in depth is served as baseline)

Figure 9.1 shows the absolute and relative differences of AP and RP values when different *qrels* are used. Every data point in Figure 9.1 is the average of all submitted runs in all year groups. One general tendency for the two measures is: the shallower the pool is, the larger the difference is. However, AP is the worst considering the difference rate. When using a pool of 10 documents in depth, the absolute difference rate is as large as 44% and the relative difference rate is 31% for AP. In the same condition, they are 32% and 10% for RP. In all the cases, relative difference is smaller than corresponding absolute difference. In addition, similar conclusions are observed for NDCG and NAPD. The difference rates for them are close to that for RP, but are lower than that for AP.

Next we explore the relation of pool depth and percentage of relevant documents identified. The result is shown in Figure 9.2. The curve increases quickly at the beginning and then slows down, but keeps increasing when the pool depth reaches 100. From the curve's tendency, it seems that the increase will continue for sometime. Using some curve estimation techniques as used by Zobel [132], we find that very likely 20% - 40% of the relevant documents can be found if the pool expands from the point of 100 documents in pool depth. From these observations, we can derive that the values of all these four measures are overestimated using a pool of 100 documents compared with their actual values when complete relevance judgment is available.

Furthermore, we investigate the impact of the number of identified relevant documents on these measures. For all 699 topics (queries) in 9 year groups, we divide them into 11 groups according to the number of relevant documents identified for them. Group 1 (G_1) comprises those topics with fewer than 10 relevant documents,

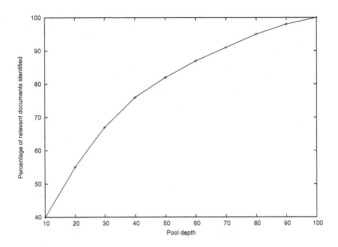

Fig. 9.2 Percentage of relevant documents identified when the pool varies in depth (the pool of 100 documents served as baseline, 100%)

group 2 (G_2) comprises those topics with 10 $\sim$ 19 relevant documents,..., group 11 (G_{11}) comprises those topics with 100 or more relevant documents. The number of topics in each group is as follows:

G_1	G_2	G_3	G_4	G_5	G_6	G_7	G_8	G_9	G_{10}	G_{11}	Total
47	16	79	76	49	33	39	27	25	17	165	699

For all these topic groups G_1 - G_{11}, we calculate the value differences of the same measure using the pools of different depths. One common tendency for the four measures is: the fewer the relevant documents are identified, the less difference the values of the same measure have with pools of different depths. For example, the value difference in G1 is always smaller than that in all other groups, while the differences in two groups G10 and G11 are larger than that of all other groups. Comparing the difference across different measures, we can observe that larger difference occurs for the measure of AP. For groups G10 and G11, the value differences of AP are 0.93% and 0.84% between the pool of 10 documents and the pool of 100 documents, while the figures for RP are 0.48% and 0.52%, respectively. From this experiment, we find that the error rate of the estimated values for any of the four measures depends on the percentage of relevant documents identified for that topic. The bigger percentage of relevant documents identified for a topic, the more accurate the estimated values for that topic. However, the numbers of relevant documents vary considerably from one topic to another. In TREC, some topics are much harder than the others and it is more difficult for information retrieval systems to catch relevant documents. Another reason is that some topics have more relevant documents than some other topics in the document collection. The numbers of relevant documents may differ greatly from one topic to another: from 1 or 2 to several hundreds. All these have considerable impact on the percentage of relevant documents identified for a given topic by the TREC pooling method. Therefore, AP, RP, NDCG, and NAPD values obtained with a pool of certain depth are not comparable across topics.

Let us see an example to explain this further. Suppose that A and B are two systems under evaluation among a group of other systems. For simplicity, we only consider 2 queries. However, the same conclusion can be drawn if more queries are used to test their effectiveness. The results are as follows:

System (query)	Observed AP	Rate of exaggeration	Real AP
A (q_1)	0.32	80%	0.32/(1+0.8)=0.1778
B (q_1)	0.25	80%	0.25/(1+0.8)=0.1389
A (q_2)	0.45	20%	0.45/(1+0.2)=0.3750
B (q_2)	0.50	20%	0.50/(1+0.2)=0.4167

According to the observed AP values, we may conclude that A is better than B, because A's AP value over two queries (0.32 + 0.45)/2 = 0.385 is greater than B's AP value over two queries (0.25 + 0.50)/2 = 0.375. However, because Query 1's AP value is overestimated by 80% and Query 2's AP value is overestimated by 20%, a modification is needed for these AP values. After that, we find that System A ((0.1778 + 0.3750)/2 = 0.2764) is worse than System B ((0.1389 + 0.4167)/2 = 0.2778). This example demonstrates that averaging the values may not be the best solution for ranking a group of retrieval systems over a group of queries in such a condition. In next subsection, we will discuss some alternatives for such a task.

9.1.2 Other Options Than Averaging All the Values for Ranking Retrieval Systems

Suppose for a certain collection of documents, we have a group of systems $IR = (ir_1, ir_2,..., ir_n)$ and a group of queries $Q = (q_1, q_2,..., q_m)$, and every system returns a ranked list of documents for every query. Now the task is to rank these systems based on their performances by any metric over these queries. If complete relevance judgment is applied, then averaging these values over all the queries is no doubt the best solution. Under partial relevance judgment, the estimated values are far from accurate and are not comparable across queries, as we have demonstrated in Section 9.1.1. Considering in a single query, if System A is better than System B with partial relevance judgment, then the same conclusion is very likely to be true with complete relevance judgment, though the difference in quantity may not be accurate. In such a situation, we may regard that these systems are involved in a number of competition events, each of which is via a query. Then the task becomes how to rank these systems according to all these m competition events. Some voting procedures such as the Borda count [3] and the Condorcet voting [61] in political science can be used here.

Both Borda count and Condorcet voting (See Section 1.2) only consider the ranks of all involved systems, but not the score values. Another option is to linearly normalize the values of a set of systems in every query into the range of [0,1], which will be referred to as the zero-one normalization method. Using this method, for every query, the top-ranked system is normalized to 1, the bottom-ranked system is normalized to 0, and all other systems are linearly normalized to a value between 0 and 1 accordingly. Thus every query is in an equal position to make contributions for the final ranking. Then all systems can be ranked according to their total scores.

9.1.3   Evaluation of the Four Ranking Methods

In this subsection we present some experimental results on the evaluation of the four methods. As in Section 9.1.1, 9 groups of runs submitted to TREC were used. For all the submissions in one year group, we calculated their effectiveness for every query using different measures. Then different ranking methods, Borda count, Condorcet voting, the zero-one normalization method, and the averaging method, were used to rank them. For these rankings obtained using different methods, we calculated Kendall's tau coefficient for each pair of rankings obtained using the same measure but different ranking method. Table 9.2 shows the results, each of which is for one of the four measures.

From Table 9.2, we can observe that Kendall's tau coefficients in all cases are quite big. For any pair in any year group, the average is always bigger than 0.8. Considering all single cases, the coefficients are less than 0.7 only occasionally. We also observe that for all the measures, the rankings from the averaging method and that from the zero-one normalization method always have the strongest correlation. This demonstrates that the averaging method and the zero-one normalization method are more similar with each other than any other pairs. In addition, the rankings from Borda count are strongly correlated with the rankings from either the averaging method or the zero-one normalization method as well. On the other hand, the correlations between the rankings from Condorcet voting and any others are always the weakest. This demonstrates that Condorcet voting is quite different from the other three methods.

Table 9.2 Kendall's tau coefficients of rankings generated by different methods using different measures (A: averaging, B: Borda, C: Condorcet, Z: zero-one)

Metric	A-B	A-C	A-Z	B-C	B-Z	C-Z
AP	0.8798	0.8143	0.9337	0.8361	0.9173	0.8308
RP	0.9072	0.8276	0.9384	0.8480	0.9379	0.8435
NAPD	0.9316	0.8416	0.9703	0.8472	0.9416	0.8445
NDCG	0.9327	0.8503	0.9692	0.8567	0.9400	0.8556

Table 9.3 Kendall's tau coefficients for AP (figures in parentheses indicate the significance level of difference compared with the averaging method)

	Averaging	Borda	Condorcet	Zeroone
1/5 all	0.7624	0.7855(.000)	0.7033(.000)	0.7765(.000)
2/5 all	0.8476	0.8658(.000)	0.7771(.000)	0.8597(.000)
3/5 all	0.8961	0.9115(.000)	0.8281(.000)	0.9071(.000)
4/5 all	0.9378	0.9454(.000)	0.8622(.000)	0.9438(.000)
Average	0.8610	0.8771[+1.87%]	0.7927[-7.93%]	0.8718[+1.25%]

Table 9.4 Kendall's tau coefficients for RP (figures in parentheses indicate the significance level of difference compared with the averaging method)

	Averaging	Borda	Condorcet	zero-one
1/5 all	0.7332	0.7418(.000)	0.6501(.000)	0.7367(.000)
2/5 all	0.8308	0.8401(.000)	0.7534(.000)	0.8387(.000)
3/5 all	0.8860	0.8943(.000)	0.8036(.000)	0.8912(.000)
4/5 all	0.9283	0.9329(.001)	0.8484(.000)	0.9311(.011)
Average	0.8446	0.8523[0.91%]	0.7639[-9.55%]	0.8494[0.57%]

Table 9.5 Kendall's tau coefficients for NAPD (figures in parentheses indicate the significance level of difference compared with the averaging method)

	Averaging	Borda	Condorcet	zero-one
1/5 all	0.7981	0.8031(.003)	0.7312(.000)	0.8036(.001)
2/5 all	0.8716	0.8761(.003)	0.7974(.000)	0.8758(.000)
3/5 all	0.9138	0.9193(.001)	0.8414(.000)	0.9187(.001)
4/5 all	0.9472	0.9504(.003)	0.8742(.000)	0.9507(.002)
Average	0.8816	0.8872[+0.64%]	0.8111[-8.00%]	0.8872[+0.64%]

Table 9.6 Kendall's tau coefficients for NDCG (figures in parentheses indicate the significance level of difference compared with the averaging method)

	Averaging	Borda	Condorcet	zero-one
1/5 all	0.7910	0.7980(.004)	0.7315(.000)	0.7962(.002)
2/5 all	0.8670	0.8751(.000)	0.8020(.000)	0.8722(.000)
3/5 all	0.9125	0.9177(.004)	0.8462(.000)	0.9165(.003)
4/5 all	0.9458	0.9504(.001)	0.8824(.000)	0.9494(.002)
Average	0.8791	0.8853[+0.71%]	0.8155[-7.23%]	0.8836[+0.51%]

Table 9.7 Kendall's tau coefficients for all the four measures when comparing the two rankings, one of which is generated with a pool of 100 documents, the other is generated with a shallow pool of 10-90 documents

	Averaging	Borda	Condorcet	zero-one
AP	0.6607	0.6800	0.4855	0.6771
RP	0.6309	0.6568	0.4851	0.6550
NAPD	0.7095	0.7167	0.5267	0.7134
NDCG	0.6981	0.7107	0.5013	0.7077

Another thing we can do is to compare those Kendall's tau coefficients using the same ranking method but different measures. Using NDCG, all Kendall's tau coefficients are the biggest (0.9006 on average). NDCG is followed by NAPD (0.8961), while AP and RP are at the bottom (0.8687). This indirectly suggests that NDCG is the most reliable measure, which is followed by NAPD and RP, while AP is the least reliable measure.

Next we investigate the issue of system ranking using different number of queries. For the same group of systems, we rank them using all the queries and using a subset of all the queries (1/5, 2/5, 3/5, and 4/5 of all the queries), then we compare these two rankings by calculating their Kendall's tau coefficient. Tables present the experimental results. In all the cases, a random process is used to select a subset of queries from all available queries. Every data point in these tables is the average of 20 pairs of rankings.

From Tables 9.3-9.6, we can see that on average Borda count and the zero-one method are the most reliable methods, the averaging method is in the middle, and Condorcet voting is the least reliable method. The difference between Condorcet voting and the others is large, while the three others are much closer with each other in performance. Although the differences between the averaging method and Borda, and between the averaging method and zero-one, are small, the differences are always significant for all four measures. Condorcet is worse than all three others at a significance level of .000. In some cases, the differences between Borda count and the zero-one method are not significant.

Finally we conducted an experiment to compare the rankings using different pools. One ranking was generated with the pool of 100 documents, and the other ranking was generated with a shallower pool of less than 100 documents. In the shallower pool, each query might be assigned a different pool depth, which was decided by a random process to choose a number from 10, 20,..., 80, and 90. The results are shown in Table 9.7, which is the average of 9 year groups, and 20 runs were performed for each year group. Again, we can observe that Condorcet is the worst, Borda count and the zero-one method is slightly better than the averaging method.

9.1.4   Remarks

As we have seen, in the condition of partial relevance judgment the averaging method may be questionable, since the values of system-oriented measures obtained from different queries are not quite comparable cross multiple queries. Several alternative methods including Borda count, Condorcet voting, and the zero-one normalization method are investigated. Our experimental results suggest that Borda count and the zero-one normalization method are slightly better than the averaging method, while Condorcet is the worst in these four methods.

The experimental results also demonstrate that with partial relevance judgment, the evaluated results can be significantly different from the results with complete relevance judgment: from their values on a system-oriented measure to the rankings of a group of information retrieval systems based on such values. Therefore, when conducting an evaluation with partial relevance judgment, we need to be careful about the results and conclusions.

9.2   Ranking Information Retrieval Systems without Relevance Judgment

In Section 9.1, we discussed some alternative approaches that can be used for ranking retrieval systems in the condition of partial relevance judgment. In this section, we go a step further. We are going to see if we can do any retrieval system ranking at all if no relevance judgment is involved. Obviously, such a undertaking is meaningful if there is any success, because human relevance judgment is very time-consuming and costly.

As a matter of fact, there has been some research on this issue. Probably one of the earliest papers is published by Soboroff et. al. [88]. Their method is a variant of the official TREC evaluation process.

The official TREC evaluation process for the ad hoc task and some other tasks is as follow.

- A group of queries are generated and released to the participants. Then participants create their queries accordingly and run their information retrieval systems to search the document collection.
- Each participant submits some number of runs. In those runs, up to 1000 top-ranked documents are included in the resultant list for each query. A subset of those runs are specified as priority runs.
- For each query, top 100 documents from those priority runs and top 10 documents from non-priority runs are collected and put into a pool. Duplicates are removed.
- For all those documents in the pool, their relevance to the query are judged by human referees. For all those documents that are not in the pool, they are not judged and therefore, regarded as irrelevant documents.
- With generated relevance judgment document in the last step, all the runs are evaluated.

Instead of checking all the documents in the pool, the method proposed by Sobo-roff et. al. [88] randomly selects a certain number of documents from the pool and assumes that all those selected are relevant documents. Thus no effort of relevance judgment is needed. This method will be referred to as the random selection method.

Another method is proposed by Wu and Crestani in [115]. Their method is based on a measure "reference count". Suppose we have a given query and a number of information retrieval systems on a common document collection. Taking a certain number of top-ranked documents returned by a retrieval system, we sum up the occurrences of these documents in the results of all other retrieval systems. In such a way, each retrieval system obtains a score, the reference count, for this particular query.

For example, suppose we have five retrieval systems ir_i $(1 \leq i \leq 5)$ and a query q, each system ir_i returns a list of documents L_i $(1 \leq i \leq 5)$ for that query. Each L_i includes 1000 documents $(d_{i,1}, d_{i,2},..., d_{i,1000})$. Let us consider L_1. For any $d_{1,j}$ $(1 \leq j \leq 1000)$, we count its occurrences $o(d_{1,j})$ in all other document lists $(L_2, L_3, L_4,$ and $L_5)$. Here we call $d_{1,j}$ the original document, while its counterparts in all other document lists are called reference documents (of $d_{i,j}$). If $d_{1,1}$ appears in L_2 and L_4, but not in L_3 and L_5, then $o(d_{1,1}) = 2$. We make a sum

$$S_1(1000) = \sum_{j=1}^{1000} o(d_{1,j})$$

which we call L_1's total reference count to the given query q. We could do the same for other lists as well. A general situation is considering the top n documents. For example, for L_i, reference count of the top 100 documents is

$$S_i(100) = \sum_{j=1}^{100} o(d_{i,j}) \qquad (9.1)$$

for $(1 \leq i \leq 5)$.

Based on the reference count obtained for each retrieval system following the above-mentioned process, we can rank these retrieval systems. This is the basic style of the reference count method. In the basic method, when calculating the reference count $o(d_{i,j})$ of $d_{i,j}$, we always add 1 to $o(d_{i,j})$ for any appearance of $d_{i,j}$'s reference document, without distinguishing the position in which that reference document appears. For example, let us suppose $d_{1,1}$ is an original document, $d_{2,1000}$ and $d_{3,1000}$ are two of its reference documents. $d_{1,2}$ is another original document, with two reference documents $d_{2,1}$ and $d_{3,1}$. Note that $d_{2,1000}$ and $d_{3,1000}$ appear at the very ends of L_2 and L_3, and $d_{2,1}$ and $d_{3,1}$ appear at the tops of L_2 and L_3. It strongly suggests that $d_{1,2}$ is more likely to be relevant to the query than $d_{1,1}$, the basic method just ignores the difference and we have $o(d_{1,1}) = o(d_{1,2}) = 2$. Since all results submitted to TREC are ranked document lists, non-discrimination is obviously not a good solution. The improvement could be addressed in two aspects: to differentiate the positions of reference documents or the position of the original document or both. The basic reference count method or any of its variants is referred to as the reference count method in general.

Both the random selection method and the reference count method work reasonably well. The rankings generated by them and the official rankings generated by NIST[1] are positively correlated. With a few collections of runs submitted to TREC (TRECs 3, 5, 6, 7, 8, and 2001), correlation coefficients (Spearman rank coefficients or Kendall Tau correlation coefficient) are between 0.3 and 0.6 for various types of metrics most of the time.

Interestingly, data fusion can also be used to deal with this problem. In the following, let us see how it works.

9.2.1 Methodology

The data fusion-based method [67] works as follows: first, from all available n runs, we select m ($m \leq n$) runs for fusion. Data fusion methods including the Borda count and the Condorcet fusion are used. In the fused result, we assume that the top k documents are relevant. Thus retrieval evaluation can be carried out by using the pseudo-quels generated.

How to choose a subset of information retrieval systems from all available ones is an issue that needs to be considered. Obviously, one straightforward way is to use all available runs. But a biased approach is also possible. In this approach, information retrieval systems that behave differently from the norm or majority of all systems are selected. First, cosine similarity between two vectors v and w can be defined as

$$s(u.v) = \frac{\sum u_i * v_i}{\sqrt{\sum (u_i)^2 * \sum (v_i)^2}}$$

The bias between these two vectors is defined as [62]

$$b(u,v) = 1 - s(u,v)$$

Two variant metrics of bias, one of which ignores the order of the documents, and the other takes account of the order of the retrieval resultant list, are provided in [62]. For the first variant, position is not considered and frequencies of documents occurrence are used as elements of the vector. For the second variant, documents at different positions are assigned different weights. More specifically, for any document d_{ij} in result L_i, $|L_i|/j$ is used to decide the weight of d_{ij}. Here $|L_i|$ is the total number of documents in result L_i and t is the ranking position of d_j. Then weighted frequencies of documents occurrence are used as elements of the vector. The norm of a group of results are defined by the collective behaviour of them. For example, if

---

[1] The National Institute of Standard and Technology is the organization that holds the annual TREC event.

$|L_i| = 12$, then the documents at ranking positions 1, 2, and 3 are assigned a weight of 12 (12/1), 6 (12/2), 4 (12/3), respectively,

Example 9.2.1. Suppose two information retrieval systems A and B are used to define the norm, and three queries are processed by each of them. Furthermore, suppose the documents retrieved by A and B, respectively, for the three queries are:

$$
A = \begin{bmatrix} a\ b\ c\ d \\ b\ a\ c\ d \\ a\ b\ c\ e \end{bmatrix} \quad B = \begin{bmatrix} b\ f\ c\ e \\ b\ c\ f\ g \\ c\ f\ g\ e \end{bmatrix}
$$

In the above matrics each row presents the documents for each query, ordered from left to right. The seven distinctive documents retrieval by either A or B are a, b, c, d, e, f, and g. Therefore, the vectors for A, B, and the norm are $v_A = (3, 3, 3, 2, 1, 0, 0)$, $v_B = (0, 2, 3, 0, 2, 3, 2)$, and $v_{norm} = v_A + v_b = (3, 5, 6, 2, 3, 3, 2)$, respectively. The similarity of v_A to v_{norm} is 0.8841, The similarity of v_B to v_{norm} is 0.8758. The bias value of A to the norm is $1-0.8841 = 0.1159$, and the bias value of B to the norm is $1-0.8758 = 0.1242$. □

If document position is considered, then $v_A = (10, 8, 4, 2, 1, 0, 0)$, $v_B = (0, 8, 22/3, 0, 2, 8/3, 7/3)$, and $v_{norm} = v_A + v_b = (10, 16, 34/3, 2, 3, 8/3, 7/3)$. The bias of A is 0.087 and the bias of B is 0.1226.

9.2.2 Experimental Settings and Results

Experiments with 4 groups of runs submitted to TREC (TREC 3, 5, 6, 7, ad hoc track) are conducted [67]. Three data fusion methods including the Borda count, the Condorcet Fusion, and CombSum (scores are generated by the reciprocal function of the rank position of the document). Also refer to Section 3.3.3 for score normalization using the reciprocal function. For calculating the bias, rank position is taken into account. After calculating the bias of each information retrieval system available, then a certain percentage of systems with the largest bias are used to do the data fusion. 10%, 20%, 30%, 40%, and 50% of the total number of systems are tried. The pool depth b can be set to different values. In the experiment, $b = 10, 20$, and 30 are tested. It is found that among three different data fusion methods, the Condorcet fusion is the best. Table 9.8 shows the result of using the method of Condorcet fusion. For Condorcet-bias, each data point is the average of using 5 different numbers of systems (10%, 20%, 30%, 40%, and 50%).

In order to make a comparison, the results of two other methods, RC and RS, are also shown in Table 9.8. The results of these two methods are taken from [115]. Three different metrics including AP, RP, and P@100, are used. From Table 9.8, we can see that Condorcet-bias is the best performer, Condorcet-normal is not as good as Condorcet-bias. But both Condorcet-normal and Condorcet-bias are better than both RS and RC considerably.

Table 9.8 Spearman rank coefficients of two rankings of all the systems submitted to TREC, one of which is generated with the automatic method, the other is generated officially by TREC (RC denotes the reference count method, RS denotes the random selection method)

Group	RC_{AP}	RS_{AP}	RC_{RP}	RS_{RP}	$RC_{P@100}$	$RS_{P@100}$	Condorcet-Normal$_{AP}$			Condorcet-Bias$_{AP}$		
							b=10	b=20	b=30	b=10	b=20	b=30
3	0.587	0.627	0.636	0.613	0.642	0.624	0.575	0.610	0.625	0.862	0.865	0.867
5	0.421	0.429	0.430	0.411	0.465	0.444	0.475	0.496	0.497	0.508	0.561	0.493
6	0.384	0.436	0.498	0.438	0.546	0.497	0.617	0.608	0.606	0.737	0.719	0.717
7	0.382	0.411	0.504	0.466	0.579	0.524	0.519	0.543	0.549	0.399	0.436	0.453
Average	0.444	0.476	0.517	0.482	0.558	0.522	0.547	0.564	0.569	0.627	0.645	0.633

9.3 Applying the Data Fusion Technique to Blog Opinion Retrieval

In recent years, blogs have been very popular on the web as a grassroots publishing platform. There are a large number of them and they cover many different aspects of people's life. A blog may be owned by an individual or a company. Posts on events, opinions, and so on, can be published by the owner and comments can be made by the readers accordingly.

A lot of research has been conducted on blogging systems and some related issues, such as blog search engines, notification mechanisms, detecting splogs, facet-based opinion retrieval, and so on, are discussed in [20, 25, 40, 43, 52, 92, 97] and others. Opinion retrieval is one of the important issues addressed in blogging systems. Usually, an opinion-finding system is built on the top of a conventional information retrieval system, which retrieves a list of relevant documents with scores for a given query, but not care if they are opinionated or not. Then an opinion finding subsystem is used to score all the documents retrieved. Finally, all the retrieved documents are re-ranked using the combination of scores obtained at both stages. Several different approaches for the opinion finding subsystem are investigated in [39, 97, 128, 130], among others.

In a sense, blog opinion retrieval systems are more complicated than conventional information retrieval systems, and many different kinds of techniques can be used together in any individual blog opinion retrieval system. In such a scenario, we hypothesize that the data fusion technique is very likely a useful technique for blog opinion retrieval. In this section, we empirically investigate this issue by extensive experimentation.

In 2006, TREC introduced the blog retrieval track. At first, only the opinion finding task was carried out. In the following three years, polarity opinion finding and distillation tasks had been added. For each of those tasks, dozens, even hundreds of runs were submitted for evaluation. This provide us a very good benchmark to test all sorts of systems, techniques, and so on. Especially, in the TREC 2008 opinion finding task, a total of 191 runs were submitted from 19 groups [69], and each of them includes a ranked list of up to 1000 post documents for each of a total of 150 queries. Since the number of runs submitted and the number of queries used are large, it is a very good data set for us to test the data fusion technique [107].

9.3.1   Data Set and Experimental Settings

In the TREC 2008 blog track, "Blog06" test collection was used. The summary information is shown in Table 9.9 [68].

Table 9.9 Summary information of the "Blog06" test collection and its corresponding statistics

Quantity	Value
Number of Unique Blogs	100,649
RSS	62%
Atom First Feed Crawl Last Feed Crawl 21/02/2006	
Number of Feeds Fetches	753,681
Number of Permalinks	3,215,171
Number of Homepages	324,880
Total Compressed Size	25GB
Total Uncompressed Size	148GB
Feeds (Uncompressed)	38.6GB
Permalinks (Uncompressed)	88.8GB
Homepages (Uncompressed)	20.8GB

Opinion retrieval is one of the tasks in the blog track. It is used to locate blog posts that express an opinion about a given target. A target can range from the name of a person or organization to a type of technology, a new product, or an event. A total of 150 topics (851-950, 1001-1050) were used in the 2008 Blog track. Among them, 50 (1001-1050) were new ones, 50 (851-900) were used in 2006 and 50 (901-950) were used in 2007. An example of a topic is shown below.

> Topic 1001
> Description: Find opinions of people who have sold a car, purchased a car, or both, through Carmax.
> Narrative: Relevant documents will include experiences from people who have bought or sold a car through Carmax and expressed an opinion about the experience. Do not include posts where people obtain estimates from Carmax but do not buy or sell an auto with Carmax.

5 standard baselines were provided by NIST (National Institute of Standards and Technology, holder of the TREC workshops) for the 2008 Blog track. Information about them can be found in [69]. Then, based on any of the baselines provided, the participants can submit their final runs. 19 groups submitted a total of 191 runs to the opinion-finding task.

Each submitted run consists of up to 1000 retrieved documents for each topic. The retrieval units are the documents from the permalinks component of the Blog06 test collection. The content of a blog is defined as the content of the post itself and all the comments to the post. Analogous to other TREC tracks, the blog track uses the pool policy for retrieval evaluation: a pool was formed from the submitted runs of the participants. The two highest priority runs per group were pooled to depth 100. The remaining runs were pooled to depth 10. Only those documents in the pool are judged.

In the experiment, three score normalization methods are used: Borda, the fitting linear score normalization method, and the binary logistic regression model. The data fusion methods involved are: CombSum, CombMNZ, the linear combination method with performance level weighting (LCP), performance square weighting (LCP2), and weights obtained by multiple linear regression (LCR). All combinations of different score normalization methods and data fusion methods were tested. Such a diversified combination of data fusion methods and score normalization methods is helpful for us to have a good observation of the achievement of the technologies available.

We divide all 150 topics into three groups of equal size. Topics 851, 854,..., 950, 1003, ..., 1048 are in group 1, topics 852, 855,..., 948, 1001, ..., 1049 are in group 2, and all the rest are in group 3. One group (1, or 2, or 3) is used as training data to decide the weights for the linear combination method, two other groups (2 and 3, or 1 and 3, or 1 and 2) are used as test data.

For all the data fusion methods involved, we randomly selected 5, 10, 15,..., 60 component systems from all available ones to test their effectiveness. For any given number, 200 combinations were tested.

9.3.2 Experimental Results

Four metrics are used for retrieval evaluation. They are: average precision over all relevant documents (AP), recall-level precision (RP), precision at 10 document level (P@10), and reciprocal rank (RR).

In these four metrics, AP and RP, and P@10 and RR show a closer correlation than other pairs (AP and P@10, AP and RR, RP and P@10, and RP and RR). This can be seen from the definitions of these four metrics. We also look at them from the fusion results with all 5 different fusion methods, 3 different score normalization methods, and different number (5, 10, ..., 60) of component systems. Kendall's tau_b and Spearman's rho coefficients are calculated, which are shown in Table 9.10. We also calculate Kendall's tau_b and Spearman's rho coefficients of all component systems, which are shown in Table 9.11.

From Table 9.10, we can see that Kendall's correlation coefficient between AP and RP is .819, and the correlation coefficient between P@10 and RR is .818, while the four others are .708, .659, .670, and .574, which are lower than .819 and .818 significantly. In Table 9.10, we can obtain very similar observations for Spearman's

Table 9.10 Kendall's tau_b and Spearman's rho coefficients of all four metrics for all fused results

	AP	RP	P@10	RR
AP	.819 & .922	.708 & .875	.659 & .819	
RP		.670 & .840	.574 & .741	
P@10			.818 & .938	

Table 9.11 Kendall's tau_b and Spearman's rho coefficients of all four metrics for all component results in parentheses

	AP	RP	P@10	RR
AP	.982 & .990	.851 & .788	.802 & .683	
RP		.911 & .793	.874 & .663	
P@10			.982 & .874	

coefficients, though all correlation coefficients are closer to 1 than their counterparts – Kendall's correlation coefficient. The Spearman's coefficients between AP and RP, and between P@10 and RR, are .922 and .938, while the four others are .875, .819, .840, and .741. Therefore, in the following, when we present experimental results, we are mainly focused on AP and RR and provide detailed data for both of them, while only brief information will be given for RP and P@10.

Table 9.12 Performance (AP) of all data fusion methods (Borda normalization; LCP, LCP2, and LCR denote the linear combination method with performance level weighting, performance square weighting, and weights decided by multiple linear regression, respectively; figures with "+" or "-" indicate they are different (better or worse) from the best component system significantly at a confidence level of 95%; figures in bold are the best in that line)

Number	Best	CombSum	CombMNZ	LCP	LCP2	LCR
5	0.3779	0.4157^+	0.4098^+	0.4248^+	0.4279^+	$\mathbf{0.4410^+}$
10	0.4032	0.4473^+	0.4380^+	0.4541^+	$\mathbf{0.4577^+}$	0.4542^+
15	0.4165	0.4576^+	0.4469^+	0.4635^+	$\mathbf{0.4673^+}$	0.4643^+
20	0.4310	0.4672^+	0.4557^+	0.4724^+	0.4758^+	$\mathbf{0.4775^+}$
25	0.4423	0.4720^+	0.4598^+	0.4773^+	0.4814^+	$\mathbf{0.4836^+}$
30	0.4488	0.4732^+	0.4614^+	0.4784^+	0.4826^+	$\mathbf{0.4899^+}$
35	0.4506	0.4756^+	0.4639^+	0.4798^+	0.4834^+	$\mathbf{0.4889^+}$
40	0.4599	0.4752^+	0.4632	0.4800^+	0.4843^+	$\mathbf{0.4898^+}$
45	0.4674	0.4783^+	0.4663	0.4829^+	0.4871^+	$\mathbf{0.4949^+}$
50	0.4706	0.4804^+	0.4683	0.4847^+	0.4888^+	$\mathbf{0.4966^+}$
55	0.4730	0.4799	0.4678	0.4843^+	0.4886^+	$\mathbf{0.4941^+}$
60	0.4806	0.4807	0.4687^-	0.4856^+	0.4905^+	$\mathbf{0.5012^+}$
Average	0.4435	0.4669	0.4558	0.4723	0.4764	**0.4813**
		(5.28%)	(2.77%)	(6.70%)	(7.42%)	(8.52%)

Table 9.13 Performance (AP) of all data fusion methods (the fit linear normalization)

Number	Best	CombSum	CombMNZ	LCP	LCP2	LCR
5	0.3779	0.4168$^+$	0.4123$^+$	0.4227$^+$	0.4247$^+$	**0.4254$^+$**
10	0.4032	0.4470$^+$	0.4398$^+$	0.4506$^+$	0.4529$^+$	**0.4593$^+$**
15	0.4165	0.4552$^+$	0.4483$^+$	0.4584$^+$	0.4611$^+$	**0.4726$^+$**
20	0.4310	0.4622$^+$	0.4555$^+$	0.4649$^+$	0.4679$^+$	**0.4880$^+$**
25	0.4423	0.4669$^+$	0.4598$^+$	0.4697$^+$	0.4731$^+$	**0.4934$^+$**
30	0.4488	0.4668$^+$	0.4603$^+$	0.4697$^+$	0.4734$^+$	**0.4983$^+$**
35	0.4506	0.4683$^+$	0.4619$^+$	0.4702$^+$	0.4732$^+$	**0.4966$^+$**
40	0.4599	0.4680$^+$	0.4615	0.4705$^+$	0.4741$^+$	**0.5014$^+$**
45	0.4674	0.4711$^+$	0.4644	0.4730$^+$	0.4766$^+$	**0.5061$^+$**
50	0.4706	0.4720	0.4657$^-$	0.4738	0.4771$^+$	**0.5073$^+$**
55	0.4730	0.4721	0.4656$^-$	0.4740	0.4775	**0.5083$^+$**
60	0.4806	0.4728$^-$	0.4661$^-$	0.4749$^-$	0.4787	**0.5112$^+$**
Average	0.4435	0.4616	0.4551	0.4643	0.4675	**0.4890**
		(4.08%)	(2.62%)	(4.69%)	(5.41%)	(10.26%)

Table 9.14 Performance (AP) of all data fusion methods (multiple linear regression)

Number	Best	CombSum	CombMNZ	LCP	LCP2	LCR
5	0.3779	0.4171$^+$	0.4175$^+$	0.4248$^+$	**0.4269$^+$**	0.4258$^+$
10	0.4032	0.4419$^+$	0.4410$^+$	0.4475$^+$	0.4514$^+$	**0.4604$^+$**
15	0.4165	0.4499$^+$	0.4488$^+$	0.4545$^+$	0.4585$^+$	**0.4734$^+$**
20	0.4310	0.4556$^+$	0.4542$^+$	0.4601$^+$	0.4646$^+$	**0.4903$^+$**
25	0.4423	0.4590$^+$	0.4575$^+$	0.4637$^+$	0.4687$^+$	**0.5000$^+$**
30	0.4488	0.4600$^+$	0.4586$^+$	0.4643$^+$	0.4691$^+$	**0.5056$^+$**
35	0.4506	0.4607$^+$	0.4594$^+$	0.4647$^+$	0.4691$^+$	**0.5098$^+$**
40	0.4599	0.4607	0.4593	0.4651	0.4701$^+$	**0.5155$^+$**
45	0.4674	0.4625$^-$	0.4611$^-$	0.4666	0.4716$^+$	**0.5212$^+$**
50	0.4706	0.4637$^-$	0.4624$^-$	0.4676	0.4723	**0.5257$^+$**
55	0.4730	0.4636$^-$	0.4624$^-$	0.4675$^-$	0.4724	**0.5266$^+$**
60	0.4806	0.4646$^-$	0.4631$^-$	0.4688$^-$	0.4740$^-$	**0.5327$^+$**
Average	0.4435	0.4549	0.4538	0.4588	0.4641	**0.4989**
		(2.57%)	(2.32%)	(3.45%)	(6.45%)	(12.49%)

First let us use AP to evaluate the experimental results. Tables 9.12 to 9.14 present the experimental results with three different score normalization methods. LCP, LCP2, and LCR denote the linear combination method with performance level weighting, performance square weighting, and weights decided by multiple linear regression, respectively.

From Tables 9.12 to 9.14, we can see that, generally speaking, all data fusion methods are effective[2]. On average, all of them outperform the best component system. The smallest improvement rate over the best is 2.32% for CombMNZ with

---

[2] Each data value in Tables 9.12-9.17 and Figures 9.3-9.5 is the average of 200 randomly selected combinations $\times$ 100 queries per test set $\times$ 3 different test sets.

Table 9.15 Performance (RR) of all data fusion methods (the Borda normalization)

Number	Best	CombSum	CombMNZ	LCP	LCP2	LCR
5	0.8036	0.8330[+]	0.8323[+]	0.8451[+]	**0.8490**[+]	0.8416[+]
10	0.8236	0.8638[+]	0.8614[+]	0.8698[+]	**0.8723**[+]	0.8693[+]
15	0.8353	0.8724[+]	0.8682[+]	0.8762[+]	**0.8769**[+]	0.8746[+]
20	0.8506	0.8818[+]	0.8775[+]	0.8840[+]	**0.8846**[+]	0.8838[+]
25	0.8585	0.8840[+]	0.8799[+]	0.8860[+]	**0.8875**[+]	0.8846[+]
30	0.8626	0.8844[+]	0.8804[+]	0.8866[+]	**0.8880**[+]	0.8844[+]
35	0.8652	**0.8879**[+]	0.8847[+]	0.8871[+]	0.8878[+]	0.8849[+]
40	0.8714	0.8857[+]	0.8807[+]	0.8869[+]	**0.8881**[+]	0.8838[+]
45	0.8766	0.8883[+]	0.8834[+]	0.8879[+]	**0.8886**[+]	0.8852[+]
50	0.8804	0.8884[+]	0.8841	0.8886[+]	**0.8900**[+]	0.8868[+]
55	0.8811	**0.8886**[+]	0.8826	0.8875[+]	0.8878[+]	0.8876[+]
60	0.8881	0.8896	0.8836[−]	0.8887	**0.8899**	0.8871
Average	0.8581	0.8790	0.8754	0.8812	**0.8825**	0.8795
		(2.44%)	(2.02%)	(2.69%)	(2.84%)	(2.49%)

Table 9.16 Performance (RR) of all data fusion methods (the fit linear normalization)

Number	Best	CombSum	CombMNZ	LCP	LCP2	LCR
5	0.8036	0.8383[+]	0.8370[+]	**0.8396**[+]	0.8385[+]	0.8301[+]
10	0.8236	0.8566[+]	0.8570[+]	0.8580[+]	**0.8582**[+]	0.8576[+]
15	0.8353	0.8639[+]	0.8646[+]	0.8644[+]	0.8648[+]	**0.8701**[+]
20	0.8506	0.8710[+]	0.8734[+]	0.8712[+]	0.8725[+]	**0.8824**[+]
25	0.8585	0.8741[+]	0.8767[+]	0.8739[+]	0.8754[+]	**0.8850**[+]
30	0.8626	0.8727[+]	0.8765[+]	0.8734[+]	0.8750[+]	**0.8910**[+]
35	0.8652	0.8733[+]	0.8784[+]	0.8737[+]	0.8756[+]	**0.8858**[+]
40	0.8714	0.8741	0.8786[+]	0.8744	0.8767[+]	**0.8899**[+]
45	0.8766	0.8764	0.8811[+]	0.8762	0.8776	**0.8921**[+]
50	0.8804	0.8790	0.8839	0.8786	0.8806	**0.8912**[+]
55	0.8811	0.8790	0.8839	0.8781	0.8800	**0.8904**[+]
60	0.8881	0.8798[−]	0.8845	0.8797[−]	0.8812[−]	**0.8955**[+]
Average	0.8581	0.8699	0.8730	0.8701	0.8713	**0.8801**
		(1.38%)	(2.35%)	(1.40%)	(1.54%)	(2.56%)

binary logistic regression, and the largest improvement rate over the best is 12.49% for LCR, also with binary logistic regression. However, the number of component systems is an important factor that affects the performance of most data fusion methods (except LCR) significantly. When a small number of component systems are fused, all data fusion methods outperform the best component system by a clear margin. When the number of component systems is above a threshold, some data fusion methods become less effective than the best component system, though such a threshold varies considerably across different data fusion methods and score normalization methods. With all three normalization methods, CombMNZ is the worst, which followed by CombSum, LCP, and LCP2, while LCR is the best. The

Table 9.17 Performance (RR) of all data fusion methods (multiple linear regression)

Number	Best	CombSum	Co,bMNZ	LCP	LCP2	LCR
5	0.8036	0.8436$^+$	0.8447$^+$	0.8440$^+$	0.8427$^+$	**0.8462$^+$**
10	0.8236	0.8555$^+$	0.8579$^+$	0.8560$^+$	0.8564$^+$	**0.8693$^+$**
15	0.8353	0.8582$^+$	0.8609$^+$	0.8589$^+$	0.8604$^+$	**0.8809$^+$**
20	0.8506	0.8629$^+$	0.8657$^+$	0.8635$^+$	0.8656$^+$	**0.8958$^+$**
25	0.8585	0.8647$^+$	0.8675$^+$	0.8654$^+$	0.8686$^+$	**0.9042$^+$**
30	0.8626	0.8602	0.8640	0.8611	0.8642	**0.9073$^+$**
35	0.8652	0.8601$^-$	0.8644	0.8607	0.8641	**0.9106$^+$**
40	0.8714	0.8594$^-$	0.8633$^-$	0.8607$^-$	0.8650$^-$	**0.9145$^+$**
45	0.8766	0.8602$^-$	0.8659$^-$	0.8589$^-$	0.8651$^-$	**0.9200$^+$**
50	0.8804	0.8589$^-$	0.8647$^-$	0.8585$^-$	0.8636$^-$	**0.9225$^+$**
55	0.8811	0.8775	0.8635$^-$	0.8585$^-$	0.8640$^-$	**0.9245$^+$**
60	0.8881	0.8579$^-$	0.8635$^-$	0.8587$^-$	0.8646$^-$	**0.9284$^+$**
Average	0.8581	0.8599	0.8622	0.8587	0.8620	**0.9020**
		(0.21%)	(0.48%)	(0.07%)	(0.45%)	(5.12%)

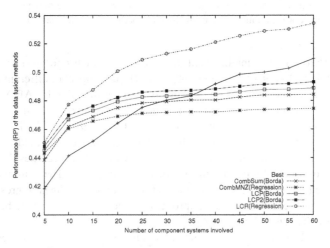

Fig. 9.3 Performance (RP) comparison of different data fusion methods for each given number of component systems (for each data fusion method, only the best performance is presented with the corresponding score normalization method that is indicated in parentheses)

combination of binary logistic regression for score normalization and multiple linear regression for system weighting is the best, which achieves an improvement rate of 12.49%. Two tailed T test is also carried out to test the significance of the difference between any data fusion method and the best component system. If the difference is significant at the level of 0.95, then a "+" or "-" sign will be put as a superscript of the corresponding value. Figures in bold are the best in that line (setting). In 33 out of a total of 36 settings, LCR is the best, while the rest 3 go to LCP2.

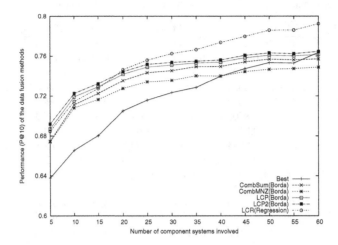

Fig. 9.4 Performance (P@10) comparison of different data fusion methods for each given number of component systems (for each data fusion method, only the best performance is presented with the corresponding score normalization method that is indicated in parentheses)

Tables 9.15 to 9.17 present the results using RR as the metric for retrieval evaluation. From Tables 9.15-9.17, we can see that for RR, the impact of different normalization methods on the data fusion methods is stronger than that for AP. When the Borda normalization is used, LCR manages an improvement rate of 2.49% over the best component system. It is slightly worse than LCP (2.69%) and LCP2 (2.84%). But LCR is still the best when two other score normalization methods are used. When the fit linear normalization method is used, CombMNZ becomes the second best data fusion method with an improvement rate of 2.35%. The combination of binary logistic regression for score normalization and multiple linear regression for system weighting achieves the best improvement rate of 5.12% over the best component system. This is in line with AP.

When using RP to evaluate these data fusion methods, the results are very similar to that of using AP. Figure 9.3 shows the result of using RP. When using P@10 to evaluate these data fusion methods, the results are similar to that of using RR. Figure 9.4 shows the results of using P@10. For both RP and P@10, all data fusion methods are better than the best component systems when a small number of results are fused. However, when a large number of results are fused, only LCR consistently outperforms the best component system. In both Figures 9.3 and 9.4, for each data fusion method, only the best result is presented with one of the three score normalization methods.

It may be arguable that the conditions for CombSum, CombMNZ, and the linear combination method are not the same; since training is required for the linear combination method to decide weights for all component systems, but no training is required for CombSum and CombMNZ. As a matter of fact, LCP and LCP2 can be regarded as special CombSum with some information (estimated performance)

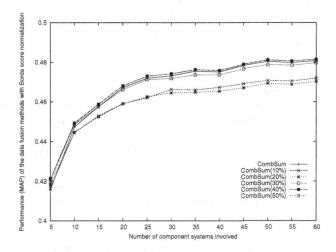

Fig. 9.5 Performance (AP) comparison of CombSum with all and a subset of good systems (CombSum($x\%$): removing those whose performance is at least $x\%$ below the average of all component systems)

available about component systems. Since the performances of all component systems can be estimated, we normalize scores of each component system into a different range that is associated with its estimated performance, then the same fusion method as CombSum is used to calculate scores for all the documents involved. Here we try another way to improve CombSum, suppose that the estimated performance of all component systems is available. The idea is to remove a number of the poorest component systems to make fusion results better. What we did is like this: for all component systems involved, we average their performances, then we remove those systems whose performance is $x\%$ below the average performance ($x = 10, 20, 30, 40,$ and 50). All settings are as before, but only the Borda score normalization method is used. The experimental result is shown in Figure 9.5.

From Figure 9.5 we can see that, when $x = 10\%$ or 20%, the fused results are worse than the fused result of CombSum(all) – all component results are involved; when $x = 30\%$ or x = 50%, the fused result is very close to that of CombSum(all); when $x = 40\%$, the fused result is slightly better than that of CombSum(all). This experimental result suggests that it is quite difficult to obtain better fusion results by removing poor component results.

In summary, one major observation from this study is: in most cases, the combination of the binary logistic regression for score normalization and multiple linear regression for weights assignment is the most effective approach, especially when a relatively large number of component systems are fused. On average, the improvement rate over the best component system is 12.49% for AP, 9.73% for RP, 5,37% for P@10, and 5.12% for RR.

Apart from the above one, we also have some other observations as follows:

1. CombSum and CombMNZ are always close. Most of the time CombSum is a little better than CombMNZ. Sometimes the difference is significant, sometimes it is not.
2. With very few exceptions, LCP2 is always a little better than LCP. The difference between them is very often significant.
3. When a relatively small number (say, 5 or 10) of component systems are fused, then all data fusion methods outperform the best component system by a clear margin.
4. For the four metrics used for retrieval evaluation, the correlation between AP and RP, and between P@10 and RR are stronger than other combinations.
5. All the data fusion methods are more effective for improvement over the best component result when the results are measured by AP or RP than by P@10 or RR.
6. For CombSum, CombMNZ, LCP, and LCP2, Borda is the best for score normalization; for LCR, logistic regression is the best for score normalization.
7. It is quite difficult for CombSum to obtain better fusion results by removing some poorest component results.

9.3.3   Discussion and Further Analysis

Most observations above are reasonable and had been seen before in this book and elsewhere, for example, in [3, 28, 48, 112, 117, 121] and others. However, one observation that has not been seen before is: the number of component systems makes significant difference for CombSum and CombMNZ. Although the performance of CombSum and CombMNZ increases with the number of component systems, its speed is slower than the best component systems. This phenomenon needs to be investigated carefully. We find that in a total of 191 component systems, a few are much better than the others. Particularly the top-one system is $NOpMMs43$, with a AP value of 0.5395, which is followed by $KGPBASE4$ (0.4800) and $KGPFILTER$ (0.4699), while the average of all systems is only 0.3083. If only a few component systems (say, fewer than 30) are chosen, then it is less likely that we can get one of the best. If a relatively large number of component systems are chosen, then we have much more chance to get one of the best. This is a very different situation from some previous experiments. For example, in [48], only 6 component systems were chosen for fusion and all of them are close in performance.

To be sure about this, we do an analysis like this: for the experiment carried out, we only choose those combinations that do not include $NOpMMs43$ – the best component system. Then we calculate those data fusion methods' performance and the best component system's performance as before. Figure 9.6 shows the result.

In Figure 9.6, we can see that both "Best" and "Best (top 1 removed)" increase with the number of component systems. However, "Best (top 1 removed)" increases at a quicker pace than "Best" does. On the other hand, presence of the best

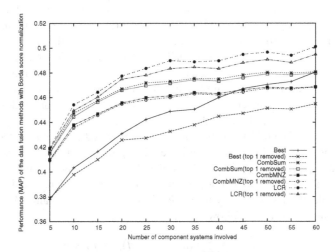

Fig. 9.6 Performance (AP) comparison of several data fusion methods with all vs with all but the best component system

component system does not affect CombSum and CombMNZ very much. "Comb-Sum" and "CombSum (top 1 removed)" are always very close to each other, so do "CombMNZ" and "CombMNZ (top 1 removed)". The difference between "LCR" and "LCR (top 1 removed)" is bigger than that between "CombSum" and "Comb-Sum (top 1 removed)", and between "CombMNZ" and "CombMNZ (top 1 removed)", but smaller than that between "Best" and "Best (top 1 removed)". This tell us that if CombSum or CombMNZ are used, then we should carefully choose a group of component systems that are more or less at the same performance level. Otherwise, the linear combination method is a better option.

9.4 TREC Systems via Combining Features and/or Components

The data fusion technique has been widely used in the implementation of various kinds of information retrieval systems to improve retrieval performance. In this section, we are going to discuss six such systems. All of them are selected from those participated in the TREC events in recent years.

9.4.1 A Hierarchical Relevance Retrieval Model for Entity Ranking

The first system we are going to discuss is a joint work by researchers in Purdue University (USA) and Kumming University of Science and Technology (China) [27].

Three runs, *KMR1PU*, *KMR2PU*, *KMR3PU*, were submitted to the TREC 2009 entity track. The goal of the entity track is to perform entity-oriented search tasks on the world wide web. It is believed that many user information needs would be better answered by specific entities instead of just any type of documents.

The system is implemented using a hierarchical relevance retrieval model for entity ranking. In this hierarchical model, three levels of relevance are examined. They are document, passage and entity. The final ranking score is a linear combination of relevance scores from all three levels.

The problem of identifying candidates who are related entities is stated as follows: what is the probability of a candidate e being the target entity given a query q and target type t? It is decided by the probability $p(e|q,t)$, and all the candidates can be ranked according to such probabilities. The top k candidates are deemed the most probable entities. By applying Bayes' Theorem and the chain rule, $p(e|q,t)$ can be decomposed into the following form

$$p(e|q,t) \propto \sum_d \sum_s p(q|d)p(q|s,d)p(e|q,t,s,d)$$

where s denotes a supporting passage in a supporting document d. The first quantity $p(q|d)$ on the right hand side is the probability that the query is generated by the supporting document, which reflects the association between the query and the document. Similarly, the second quantity $p(q|s,d)$ reflects the association between the query and the supporting passage. The last quantity $p(e|q,t,s,d)$ is the probability that candidate e is a related entity given passage s, type t and query q. In summary, this probabilistic retrieval model considers the relevance at three different levels: document, passage and entity. However, accurately estimating these probabilities is difficult for generative probabilistic language modeling techniques. Instead, motivated by the idea of the hierarchical relevance model, a linear combination of relevance scores from these three levels to yield the final ranking score $f(e|q,t)$ as follows:

$$f(e|q,t) = \sum_d \sum_s a_1 f(q|d) + a_2 f(q|s,d) + a_3 f(e|q,t,s,d) \qquad (9.2)$$

where a_1, a_2, and a_3 are coefficients for $f(q|d)$, $f(q|s,d)$ and $f(e|q,t,s,d)$, respectively. See [27] for how to calculate those scores needed. However, it is not clear what values are used for those coefficients a_1, a_2, and a_3 or how to decide their values.

The system performs very well. The performance of all three runs submitted are very close since only small changes are made for producing those different runs. *KER1PU* is top-ranked in all the runs submitted to the entity track from 13 systems [6].

9.4.2   Combination of Evidence for Effective Web Search

The second system is developed by researchers in Carnegie Mellon University [65]. The Lemur project's search engine is used. Several runs from this system were submitted to the TREC 2010 web track.

The goal of the web track is to explore and evaluate web retrieval technologies [23]. The same as 2009, TREC 2010 web track contained three tasks: ad hoc, diversity and spam filtering. The ad hoc task ranks systems according to their performance based on manual relevance assessments. For every query, a specific information need was specified. With the diversity task, the goal is to return a ranked list that provides a complete coverage of the query and avoids redundancy.

In this system, different retrieval components and their combinations are used. In the following we briefly review some of them. The results presented are the results on the training set (web track 2009) for 2010.

Firstly, priors can be added directly in the query and are treated as ordinary query terms. Different kinds of priors including PageRank, Spam, and URL, are explored. It is found that Spam is more helpful than the two others.

The index file contains title, inlink, heading, and document fields. This can be used to construct a mixture model. After tuning the parameters in combination with the priors, it is observed that the best performance is with title (0.1), inlink (0.2) and document (0.7).

The dependency model [58] is also explored. This model expands the original query with subqueries that add proximity constraints. Both ordered and unordered constraints are added. After parameter tuning, it is observed that the unordered component is not effective.

The two above-mentioned models are combined linearly by the #weight operator in Lemur. After training, it is found that the best performance is with the mixture model (0.9) and the dependency model (0.1).

In Table 9.18 the effectiveness of the mixture model, dependency model and their combination is presented. The baseline system of these runs uses the optimized priors.

Table 9.18 Results of the dependency and mixture models and their combination over all relevance assessments of 2009

Run	AP	P@10
Dependency model	0.0627	0.2560
Mixture model	0.0751	0.3120
Combination	0.0881	0.3360

Two types of query expansion are explored: Wikipedia expansion and combining Wikipedia and expansion over top retrieved web pages. For the latter type, two further strategies are explored. They are:

- Union: expand by taking the union of the top terms of both expansion term sets.
- Intersection: expand by taking the intersection for both sets and averaging the weights.

Table 9.19 Performance (AP, P@10, ERR20, and NDCG20) of the three different runs submitted to the ad hoc task

Run	2009		2010			
	AP	P@10	AP	P@10	ERR20	NDCG20
Baseline (cmuBase10)	0.0881	0.3360	0.0976	0.2833	0.0913	0.1420
Wikipedia (cmuWiki10)	0.1399	0.4520	0.1574	0.4208	0.1121	0.2118
Union (cmuFuToP@10)	0.1040	0.3320	0.1177	0.3125	0.1001	0.1583
Intsection (cmuComb10)	0.1137	0.3780	0.1209	0.3250	0.0983	0.1690

Table 9.20 Performance (ERR-IA20, NDCG20, NRBP, AP-IA) of the three different runs submitted to the diversity task in TREC 2010

Run	ERR-IA20	NDCG20	NRBP	MAP-IA
Baseline (cmuBase10)	0.2019	0.3042	0.1631	0.0498
Wikipedia (cmuWiki10)	0.2484	0.3452	0.2149	0.0926
Union (cmuFuToP@10)	0.2084	0.3091	0.1708	0.0621
Intersection (cmuComb10)	0.2151	0.3236	0.1732	0.0649

Four runs were submitted to the ad hoc task and the diversity rask of the web track in TREC 2009, respectively. Parameters were tuned using parameter sweeps with the data from the web track in 2009. The baseline is *cmuBase*10, which makes use of the priors and combines the mixture and dependency model linearly. *cmuWiki*10 is the second submission. It uses Wikipedia for query expansion. Only Wikipedia pages appearing in the top results (e.g. top 1000) are considered for expansion. In the third submission, they combines Wikipedia and expansion over the top retrieved web pages. Two further strategies are taken: union (*cmuFuToP*@10) and intersection (*cmuComb*10). Table 9.19 and 9.20 shows the results, each of which is for a different task. We can see that further improvement over the baseline is achievable for all those added components in both tasks.

9.4.3 Combining Candidate and Document Models for Expert Search

The third system was developed by researchers from the University of Amsterdam [7]. A standard language modeling approach was used and documents were ranked by their log-likelihood of being relevant to a given query. A number of runs from the system were submitted to the enterprise track in TREC 2008. Here we focus on the expert search task of the system.

For the expert search task, their aim was to experiment with a proximity-based version of the candidate model [8], to combine it with document-based models, to determine the effectiveness of query modeling, and to bring in web evidence.

Their approach to ranking candidates is as follows:

$$P(ca|q) \propto P(ca) * P(q|ca) \qquad (9.3)$$

where $P(ca)$ is the priori probability of the candidate ca being an expert, and $P(q|ca)$ is the probability of ca generating the query q. There are two different ways of estimating $P(q|ca)$. The first one is called the candidate model (Model 1), and the second one is called the document model (Model 2).

Four runs submitted are:

- *UvA08ESm1b*: Model 1 using the initial query (without expansion).
- *UvA08ESm2all*: Model 2 using expanded query models and all document search features (on top of document search run *UvA08DSall*)
- *UvA08EScomb*: linear combination of Model 1 (with weight 0.7) and Model 2 (with weight 0.3). Both models use the initial query (without expansion).
- *UvA08ESweb*: linear combination of the run *UvA08EScomb* (with weight 0.75) and the web-based variation of Model 1 (with weight 0.25). The web run uses the query model from *UvA08DSexp*.

Table 9.21 Performance of several options for the expert search task in TREC 2008

Run	#rel_ret	AP	P@5	P@10	RR
UvA08ESm1b	394	.3935	.4836	.3473	.8223
UvA08ESm2all	395	.3679	.4473	.3436	.6831
UvA08EScomb	419	.4331	.4982	.3836	.8547
UvA08ESweb	425	.4490	.5527	.3982	.8721

Table 9.21 shows that the most successful strategy is to put everything together: *UvA08ESweb* outperforms the three other runs. Interestingly, Model 1 outperforms Model 2; note that the run labeled *UvA08ESm1b* does not employ query expansion, while *UvA08ESm2all* uses features that improved performance on the document search task, including query expansion. Furthermore, we see that a combination of the two methods outperforms both models on all metrics. And finally, bringing in web evidence helps improve retrieval comparison even further (see the run labeled *UvA08ESwb*).

9.4.4 Combining Lexicon-Based Methods to Detect Opinionated Blogs

The fourth system is from Indiana University's WIDIT Lab. In 2007, they submitted a few runs to the blog track's opinion task and polarity subtask. They also submitted some runs to TREC in 2005 (hard, robust, and spam tracks) and 2006 (blog track) by using the same fusion method. The goal of the opinion task is to "uncover the public sentiment towards a given entity/target". Detecting opinionated blogs on a given topic (i.e., entity/target) involves not only retrieving topically relevant blogs, but also identifying which contain opinions about the target. The usual approach to the opinion finding task is: firstly applying a traditional information retrieval method to retrieve on-topic blogs, and then re-rank the list obtained in the first step by a component that can detect opinions. Their approach to the second step is to boost the ranks of opinionated blogs based on combined opinion scores generated by multiple opinion detection methods. The polarity subtask requires classification of the retrieved blogs into positive or negative orientation. To accomplish this, they extended an opinion detection module to generate polarity scores for polarity determination.

The initial retrieval engine was implemented by a probabilistic model (Okapi BM 25 [74]). For the on-topic retrieval optimization part, it involves re-ranking the initial retrieval results based on a set of topic-related re-ranking factors that are not used in the initial ranking of documents. The topic re-ranking factors used in the study are:

- Exact match, which considers the frequency of exact query string occurrence in documents;
- Proximity match, which considers the frequency of padded query string occurrence in documents;
- Noun phrase match, which considers the frequency of query noun phrases occurrence in documents;
- Non-rel match, which considers the frequency of non-relevant nouns and noun phrase occurrence in documents.

All the re-ranking factors are normalized by document length. The on-topic re-ranking method consists of the following three steps: firstly, compute topic re-ranking scores for top n results; secondly, partition the top n results into re-ranking groups based on the original ranking and a combination of the most influential re-ranking factors. The purpose of using re-ranking groups is to prevent excessive influence of re-ranking by preserving the effect of key ranking factors. Finally, re-rank the initial retrieval results within re-ranking groups by the combined re-ranking scores.

For the opinion detection part, five modules implemented are as follows:

- High frequency module is used to identify opinions based on common opinion terms;
- Wilson's lexicon module uses Wilson's subjectivity terms [104] to supplement the high frequency lexicon;
- Low frequency module is used to find low frequency terms for opinion evidence.
- IU module is motivated by the observation that pronouns such as "I" and "You" appear very frequently in opinionated texts.
- Opinion acronym module takes the opinion acronym lexicon into consideration.

For the fusion process, the linear combination method is used with normalized scores by the zero-one score normalization method. That is

$$g(d) = \alpha * s_{original}(d) + \beta * \sum_{i=1}^{k} w_i * s_i(d) \qquad (9.4)$$

This same equation is used in two different places. The first is to obtain optimized on-topic results and the second is to obtain opinion results.

For the former case, $s_{original}(d)$ is the score that d obtains from the initial retrieval stage, while $s_i(d)$ is the scores that d obtains from those 4 factors. For the latter case, $s_{original}(d)$ is the score that d obtains from the former case, while $s_i(d)$ is the scores that d obtains from those 5 modules. A dynamic tuning interface was also implemented for helping users to tune parameters manually and display its effect immediately.

Table 9.22 Performance (AP) of the data fusion method for the topic finding in TREC 2007

	Short Query	Long Query	Fusion
Baseline	0.3367	0.3736	0.3893 (+9.5%)
Topic re-rank	0.3889 (+15.5%)	0.4082 (+9.3%)	0.4189 (+2.6%)

Table 9.23 Performance (AP) of the data fusion method for the opinion task in TREC 2007

	Short Query	Long Query	Fusion
Baseline	0.2640	0.2817	0.2900 (+2.9%)
Topic re-rank	0.2579 (-2.4%)	0.2983 (+5.9%)	0.3305 (+7.3%)
Opinion re-rank	0.2959 (+14.7%)	0.3303 (+10.7%)	0.3343 (+1.2%)

Table 9.22 and 9.23 show the topic and opinion performance (in AP) of the re-ranked results in comparison with the original retrieval results (i.e. baseline). Topic performance is the average precision computed using only the topical relevance (i.e., document is topically relevant), whereas opinion performance is computed using the opinion relevance (i.e., document is topically relevant and opinionated). Among the three main strategies, opinion re-ranking proved most useful (15% improvement in opinion AP for short query, 11% improvement for long query over topic rerank results). Although topic re-ranking resulted in 16% improvement in topic AP (short query) and 9% improvement (long query) over the initial retrieval results (Table 9.22), it reduced the opinion AP by 2.4% for short query and showed only 5.9% improvement for long query.

In other words, topic re-ranking can hurt the opinion performance by over-boosting topically relevant documents to top ranks. The ability of an opinion re-ranking method to compensate for the adverse effect of topic re-ranking is demonstrated in the short query column of Table 9.23, where opinion re-ranking shows

15% improvement over topic re-ranking compared to 12% improvement over base-line. The fusion columns of the tables show only marginal improvements over the best non-fusion results (i.e., long query) except for the topic re-rank row in Table 9.23, which shows 7.3% improvement. This suggests the usefulness of fusion by its performance over the best component system.

9.4.5   Combining Resources to Find Answers to Biomedical Questions

The fifth selected system is the one called *NLM* [24]. Several runs from this system were submitted to the question answering task of the genomics track in 2007.

The 2007 TREC genomics track focused on answering questions gathered from working biologists. Rather than finding an exact answer, the task was to extract passages containing answers from about 160,000 full-text scientific articles published in 49 genomics-related journals.

Among several others, one fusion run, *NLMFusion*, was submitted. *NLMFusion* is the fused result from five component systems, Essie [42], Indri [57], Terrier [70], Theme [103], and EasyIR [75]. The fusion method used is CombSum. Although the performance of five component runs was not uniform (see Table 9.24), the fusion run *NLMFusion* outperformed all but the Terrier component run at all evaluation levels. The same data fusion method was used for the submission to TREC in 2005 and 2006 as well.

Table 9.24 Performance (AP) of the automatic component runs and the fusion run

System/Run	Document	Passage2	Aspect
EasyIR	0.0619	0.0133	0.0222
Essie	0.2327	0.0698	0.2249
Indri	0.2209	0.0698	0.1790
Terrier	0.3008	0.0922	0.2493
Theme	0.0568	0.0110	0.0552
NLMFusion	0.3105	0.1097	0.2494

9.4.6   Access to Legal Documents: Exact Match, Best Match, and Combinations

Just as the third system, the sixth system was developed in the University of Amsterdam. But apart from this, they are completely different systems and were developed in different research groups. Quite a few runs from this system were submitted to the TREC 2007 legal track. The focus of the legal track ad hoc evaluation is recall-oriented measures since users may wish to find all relevant documents.

Mainly Lucene (version 1.9) [54], working with a vector space model, is used as the primary search engine. A Boolean model is also used as a reference model.

Two separate indexes were created. One is the full-text index, in which the full textual content of the documents, including the meta-data tags is indexed; the other is the text-only index, in which only the text inside the tags, but not the tags is indexed. *Full-text* runs on the full-text index, while *Text-only* uses the text-only index. the third run is *SelectedTerms*. It adds more words into the original query. Those extra words are selected from the topic description part. The fourth run is a fusion run *CombiTextSelectedTerms*, it combines *Text-only* and *SelectedTerms*. Next one is *refL07B*, it works with a Boolean model. The sixth run *CombiTextRef* is also a fusion run, which combines results of *Text-only* and *refL07B*. *BoolTermsOnly*, *BoolLuceneTrans*, *BoolTermsWildcardExp*, and *BoolWildcardExpLuceneTrans* are four further runs that take different query representations. *CombiTextTerms*, *CombiTextLuceneTrans*, *CombiTextTermsWildcard*, and *CombiTextWildcardLuceneTrans* are four further fusion runs. Two component results are fused for each of them. CombSum is used in all fusion runs.

Table 9.25 Performance of a group of runs for the legal track in 2007, best scores are in bold-face

Run	AP	bpref	P@10	num_rel_ret	recall_B	est_RB
Full-text (a)	0.0878	0.3266	0.2837	3,338	0.4792	0.1448
Text-only (b)	0.0880	0.3255	0.2860	3,339	0.4835	0.1548
SelectedTerms (c)	0.0355	0.2619	0.1070	2,522	0.3173	0.0772
CombiTextSelectedTerms (b+c)	0.0846	0.3302	0.2698	3,306	0.4841	0.1447
refL07B (d)	0.0167	0.2902	0.0209	2,145	0.4864	**0.2158**
CombiTextRef (b+d)	0.1181	**0.3842**	0.3209	3,553	0.4864	**0.2158**
BoolTermsOnly (e)	0.0878	0.3274	0.2535	3,016	0.4846	0.1417
BoolLuceneTrans (f)	0.0880	0.3039	0.2535	2,200	0.4172	0.1526
BoolTermsWildcardExp (g)	0.1021	0.3321	0.3140	3,352	0.5122	0.1335
BoolWildcardExpLuceneTrans (h)	0.0915	0.3250	0.2488	2,758	0.4369	0.1555
CombiTextTerms (b+e)	0.1220	0.3615	0.3326	3,473	**0.5673**	0.1843
CombiTextLuceneTrans (b+f)	0.1191	0.3683	0.3047	3,426	0.5520	0.1908
CombiTextTermsWildcard (b+g)	**0.1264**	0.3592	**0.3465**	**3,490**	0.5627	0.1644
CombiTextWildcardLuceneTrans (b+h)	0.1191	0.3665	0.3256	3,431	0.5479	0.1976

Note: num_rel_ret denotes the number of relevant documents retrieved; recall_B denotes recall at B; est_RB denotes denotes estimated recall at B. See [95] for detailed definitions of them.

Table 9.25 shows the results. In the "Run" row, the titles of all the runs are presented. For those fusion runs, the two component systems involved are also presented. For example, *CombiTextTerms* is the fused result of *Text-only* (b) and *BoolTermsOnly* (e). Thus, we use (b+e) to denote it. All the fusion runs ourperform the corresponding component runs except *CombiTextSelectedTerms*.

CombiTextSelectedTerms is not as good as one of the component system *Text −
only* on some measures. The result is understandable, since the other component
system *SelectedTerms* is much worse than *Text-only* and CombSum treats both of
them equally. For example, the P@10 value of *Text-only* is 0.2860, the P@10 value
of *SelectedTerms* is only 0.1070, while that of the fused result is 0.2698. We should
be satisfied with this since it is quite close to the best one (0.2860) and far away
from the worst (0.1070). But generally speaking, the data fusion methods work well
for other combinations, especially the fusion run *CombiTextTermsWildcard*.

9.4.7 Summary

In this section, we have reviewed six systems that take advantage of the data fusion
technique. Those TREC systems are used for different tasks inclusing the entity
track, web track, enterprise track, blog track, genomics track, and legal track. The
fused results are effective if compared them with those component results involved.
It demonstrates that the data fusion technique is useful for various kinds of tasks.

Among all six systems, four of them use the linear combination method, while
the two others use CombSum. For those using the linear combination method, one of
them (see Section 9.4) provides an interface for users to tune the weights manually,
while it is not clear how the weights are decided in the three other systems.

References

1. Alpaydin, E.: Introduction to Machine Learning. The MIT Press (2010)
2. Amitay, E., Carmel, D., Lempel, R., Soffer, A.: Scaling IR-system evaluation using term relevance sets. In: Proceedings of the 27th Annual International ACM SIGIR Conference, Sheffield, UK, pp. 10–17 (July 2004)
3. Aslam, J.A., Montague, M.: Models for metasearch. In: Proceedings of the 24th Annual International ACM SIGIR Conference, New Orleans, Louisiana, USA, pp. 276–284 (September 2001)
4. Aslam, J.A., Yilmaz, E.: Inferring document relevance from incomplete information. In: Proceedings of the Sixteenth ACM Conference on Information and Knowledge Management, Lisbon, Portugal, pp. 633–642 (November 2007)
5. Baeza-Yates, R., Ribeiro-Neto, B.: Modern Information Retrieval. ACM Press and Addison-Wesley (1999)
6. Balog, K., de Vries, A., Serdyukov, P., Thomas, P., Westerveld, T.: Overview of the trec 2009 entity track. In: TREC (2009)
7. Balog, K., de Rijke, M.: Combining candidate and document models for expert search. In: TREC (2008)
8. Balog, K., de Rijke, M.: Non-local evidence for expert finding. In: Proceedings of the 17th ACM Conference on Information and Knowledge Management, Napa Valley, California, USA, pp. 489–498 (October 2008)
9. Bartell, B.T., Cottrell, G.W., Belew, R.K.: Automatic combination of multiple ranked retrieval systems. In: Proceedings of ACM SIGIR 1994, Dublin, Ireland, pp. 173–184 (July 1994)
10. Beitzel, S., Jensen, E., Chowdhury, A., Grossman, D., Frieder, O., Goharian, N.: On fusion of effective retrieval strategies in the same information retrieval system. Journal of the American Society of Information Science and Technology 55(10), 859–868 (2004)
11. Belkin, N.J., Cool, C., Croft, W.B., Callan, J.P.: The effect of multiple query representations on information retrieval performance. In: Proceedings of ACM SIGIR 1993, Pittsburgh, USA, pp. 339–346 (June-July 1993)
12. Belkin, N.J., Kantor, P., Fox, E.A., Shaw, J.A.: Combining evidence of multiple query representations for information retrieval. Information Processing & Management 31(3), 431–448 (1995)

13. Bookstein, A.: Informetric distributions, part 1: unified overview. Journal of the American Society for Information Science 41(5), 368–375 (1990)
14. Bookstein, A.: Informetric distributions, part 2: resilience to ambiguity. Journal of the American Society for Information Science 41(5), 376–386 (1990)
15. Buckley, C., Voorhees, E.M.: Retrieval evaluation with incomplete information. In: Proceedings of ACM SIGIR Conference, Sheffield, United Kingdom, pp. 25–32 (July 2004)
16. Callan, J.: Distributed information retrieval. In: Advances in Information Retrieval, pp. 127–150 (2000)
17. Callan, J.K., Lu, Z., Croft, W.: Searching distributed collections with inference networks. In: Proceedings of the 18th Annual International ACM SIGIR Conference, Seattle, USA, pp. 21–28 (July 1995)
18. Calvé, A.L., Savoy, J.: Database merging strategy based on logistic regression. Information Processing & Management 36(3), 341–359 (2000)
19. Chapelle, O., Metlzer, D., Zhang, Y., Grinspan, P.: Expected reciprocal rank for graded relevance. In: Proceedings of the 18th ACM Conference on Information and Knowledge Management, Hong Kong, China, pp. 621–630 (November 2009)
20. Chen, Y., Tsai, F., Chan, K.: Machine larning techniques for business blod search and mining. Expert Systems with Applications 35(3), 581–590 (2008)
21. Cochran, W.G.: Sampling techniques. Jonn Wiley & Sons, Inc. (1963)
22. Cohen, R., Katzir, L.: The generalized maximum coverage problem. Information Processing Letters 108(1), 15–22 (2008)
23. Craswell, N., Fetterly, D., Najork, M.: Microsoft research at trec 2010 web track. In: TREC (2010)
24. Demner-Fushman, D., Humphrey, S.M., Ide, N.C., Loane, R.F., Mork, J.G., Ruch, P., Ruiz, M.E., Smith, L.H., Wilbur, W.J., Aronson, A.R.: Combining resources to find answers to biomedical questions. In: TREC (2007)
25. Du, H., Wagner, C.: Learning with weblogs: enhancing cognitive and social knowledge consstruction. IEEE Transactions on Professional Communication 50(1), 1–16 (2007)
26. Egozi, O., Gabrilovich, E., Markovitch, S.: Concept-based feature generation and selection for information retrieval. In: Proceedings of the Twenty-Third AAAI Conference on Artificial Intelligence, Chicago, USA, pp. 1132–1137 (July 2008)
27. Fang, Y., Si, L., Yu, Z., Xian, Y., Xu, Y.: Entity retrieval with hierarchical relevance model, exploiting the structure of tables and learning homepage classifiers. In: TREC (2009)
28. Farah, M., Vanderpooten, D.: An outranking approach for rank aggregation in information retrieval. In: Proceedings of the 30th ACM SIGIR Conference, Amsterdam, The Netherlands, pp. 591–598 (July 2007)
29. Feige, U.: A threshold of $\ln n$ for approximating set cover. Journal of ACM 45(4), 634–652 (1998)
30. Fernández, M., Vallet, D., Castells, P.: Probabilistic Score Normalization for Rank Aggregation. In: Lalmas, M., MacFarlane, A., Rüger, S.M., Tombros, A., Tsikrika, T., Yavlinsky, A. (eds.) ECIR 2006. LNCS, vol. 3936, pp. 553–556. Springer, Heidelberg (2006)
31. Foltz, P.W., Dumais, S.T.: Personalized information delivery: an analysis of information-filtering methods. Communications of the ACM 35(12), 51–60 (1992)
32. Fox, E.A., Koushik, M.P., Shaw, J., Modlin, R., Rao, D.: Combining evidence from multiple searches. In: The First Text REtrieval Conference (TREC-1), Gaitherburg, MD, USA, pp. 319–328 (March 1993)

33. Fox, E.A., Shaw, J.: Combination of multiple searches. In: The Second Text REtrieval Conference (TREC-2), Gaitherburg, MD, USA, pp. 243–252 (August 1994)
34. Freund, J.E.: Mathematical statistics, 5th edn. Prentice-Hall, Englewood Cliffs (1992)
35. Freund, J.E., Wilson, W.J.: Regression Analysis: Statistical Modeling of a Response Variable, San Diego, California. Academic Press, London (1998)
36. Fuhr, N.: A decision-theoretic approach to database selection in networked IR. ACM Transactions on Information Systems 17(3), 229–249 (1999)
37. Gravano, L., García-Molina, H.: Generalizing gloss to vector-space database and broker hierarchies. In: Proceedings of 21st VLDB Conference, Zűrich, Switzerland, pp. 78–89 (1995)
38. Hawking, D., Thistlewaite, P.: Methods for information server selection. ACM Transactions on Information Systems 17(1), 40–76 (1999)
39. Huang, X., Croft, B.: A unified relevance model for opinion retrieval. In: Proceeding of the 18th ACM Conference on Information and Knowledge Management, Hong Kong, China, pp. 947–956 (November 2009)
40. Huang, Y., Huang, T., Huang, Y.: Applying an intelligent notification mechanism to blogging systems utilizing a genetic-based information retrieval approach. Expert Systems with Applications 37(1), 705–715 (2010)
41. Hull, D.A., Pedersen, J.O., Schüze, H.: Method combination for document filtering. In: Proceedings of the 19th Annual International ACM SIGIR Conference, Zurich, Switzerland, pp. 279–287 (August 1996)
42. Ibraev, U., Ng, K.B., Kantor, P.B.: Exploration of a geometric model of data fusion. In: Proceedings of the 2002 Annual Conference of the American Society for Information Science and Technology, Philadelphia, USA, pp. 124–129 (November 2002)
43. Kwai, R.(Fun IP), Wagner, C.: Weblogging: A study of social computing and its impact on organizations. Decision Support Systems 45(2), 242–250 (2008)
44. Järvelin, K., Kekäläinen, J.: Cumulated gain-based evaluation of IR techniques. ACM Transactions on Information Systems 20(4), 442–446 (2002)
45. Kamps, J., de Rijke, M., Sigurbjörnsson, B.: Combination Methods for Crosslingual Web Retrieval. In: Peters, C., Gey, F.C., Gonzalo, J., Müller, H., Jones, G.J.F., Kluck, M., Magnini, B., de Rijke, M., Giampiccolo, D. (eds.) CLEF 2005. LNCS, vol. 4022, pp. 856–864. Springer, Heidelberg (2006)
46. Katzer, J., McGill, M.J., Tessier, J.J., Frakes, W., DasGupta, P.: A study of the overlap among document representations. Information Technology: Research and Development 1(2), 261–274 (1982)
47. Khuller, S., Moss, A., Naor, J.: The budgeted maximum coverage problem. Information Processing Letters 70(1), 39–45 (1999)
48. Lee, J.H.: Analysis of multiple evidence combination. In: Proceedings of the 20th Annual International ACM SIGIR Conference, Philadelphia, Pennsylvania, USA, pp. 267–275 (July 1997)
49. Lemur, http://www.lemurproject.org/
50. Lillis, D., Toolan, F., Collier, R., Dunnion, J.: Probfuse: a probabilistic approach to data fusion. In: Proceedings of the 29th Annual International ACM SIGIR Conference, Seattle, Washington, USA, pp. 139–146 (August 2006)
51. Lillis, D., Zhang, L., Toolan, F., Collier, R., Leonard, D., Dunnion, J.: Estimating probabilities for effective data fusion. In: Proceeding of the 33rd International ACM SIGIR Conference on Research and Development in Information Retrieval, Geneva, Switzerland, pp. 347–354 (July 2010)
52. Lin, Y., Sundaram, H., Chy, Y., Tatemura, J., Tseng, B.: Detecting splogs via temporal dynamics using self-similarity analysis. ACM Transactions on Web 2(1), 1–35 (2008)

53. Liu, K., Yu, C., Meng, W., Rishe, N.: A statistical method for estimating the usefulness of text databases. IEEE Transactions on Knowledge and Data Engineering 14(6), 1422–1437 (2002)
54. Lucene, `http://lucene.apache.org/`
55. Manmatha, R., Rath, T., Feng, F.: Modelling score distributions for combining the outputs of search engines. In: Proceedings of the 24th Annual International ACM SIGIR Conference, New Orleans, USA, pp. 267–275 (September 2001)
56. Meng, W., Yu, C., Liu, K.: Building efficient and effective metasearch engines. ACM Computing Surveys 34(1), 48–89 (2002)
57. Metzler, D., Bruce Croft, W.: Combining the language model and inference network approaches to retrieval. Information Processing & Management 40(5), 735–750 (2004)
58. Metzler, D., Bruce Croft, W.: A markov random field model for term dependencies. In: Proceedings of the 28th Annual International ACM SIGIR Conference, Salvador, Brazil, pp. 472–479 (August 2005)
59. Montague, M.: Metasearch: Data fusion for document retrieval. Technical Report TRC2002-424, Department of Computer Science, Dartmouth College, Hanover, New Hampshire, USA (2002)
60. Montague, M., Aslam, J.A.: Relevance score normalization for metasearch. In: Proceedings of ACM CIKM Conference, Berkeley, USA, pp. 427–433 (November 2001)
61. Montague, M., Aslam, J.A.: Condorcet fusion for improved retrieval. In: Proceedings of ACM CIKM Conference, McLean, VA, USA, pp. 538–548 (November 2002)
62. Mowshowitz, A., Kawaguchi, A.: Assessing bias in search engines. Information Processing & Management 38(1), 141–156 (2002)
63. Nassar, M.O., Kanaan, G.: The factors affecting the performance of data fusion algorithms. In: Proceedings of the International Conference on Information Management and Engineering, Kuala Lumpur, Malaysia, pp. 465–470 (April 2009)
64. Ng, K.B., Kantor, P.B.: Predicting the effectiveness of naive data fusion on the basis of system characteristics. Journal of the American Society for Information Science 13(50), 1177–1189 (2000)
65. Nguyen, D., Callan, J.: Combination of evidence for effective web search. In: TREC (2010)
66. Nottelmann, H., Fuhr, N.: From retrieval status values to probabilities of relevance for advanced ir applications. Information Retrieval 6(3-4), 363–388 (2003)
67. Nuray, R., Can, F.: Automatic ranking of information retrieval systems using fusion data. Information Processing & Management 42(3), 595–614 (2006)
68. Ounis, I., de Rijke, M., Macdonald, C., Mishne, G., Soboroff, I.: Overview of the trec-2006 blog track. In: Proceeding of the 15th Text Retrieval Conference, Gaithersburg, MD, USA (2006)
69. Ounis, I., Macdonald, C., Soboroff, I.: Overview of the trec-2008 blog track. In: Proceeding of the 17th Text Retrieval Conference, Gaithersburg, MD, USA (2008)
70. Ounis, I., Amati, G., Plachouras, V., He, B., Macdonald, C., Johnson, D.: Terrier information retrieval platform. In: Advances in Information Retrieval, 27th European Conference on IR Research, ECIR, Santiago de Compostela, Spain, pp. 517–519 (March 2005)
71. Press, W.H., Teukolsky, S.A., Vettering, W.T., Flannery, B.P.: Numerical Recipes in C: the Art of Scientific Computing. Cambridge University Press, UK (1995)
72. Rao, S.S.: Optimisation Theory and Applications. Wiley Eastern Limited (1984)
73. Rasolofo, Y., Hawking, D., Savoy, J.: Result merging strategies for a current news metasearcher. Information Processing & Management, 581–609 (2003)

74. Robertson, S.E., Walker, S.: Some simple effective approximations to the 2-poisson model for probabilistic weighted retrieval. In: Bruce Croft, W., van Rijsbergen, C.J. (eds.) Proceedings of the 17th Annual International ACM-SIGIR Conference on Research and Development in Information Retrieval, (Special Issue of the SIGIR Forum), July 3-6, pp. 232–241. ACM/Springer, Dublin, Ireland (1994)

75. Ruiz, M.E., Sun, Y., Wang, J., Liu, H.: Exploring traits of adjectives to predict polarity opinion in blogs and semantic filters in genomics. In: Proceedings of The Sixteenth Text REtrieval Conference (2007)

76. Saari, D.G.: Basic Geometry of Voting. Springer (1995)

77. Sakai, T., Kando, N.: On information retrieval metrics designed for evaluation with incomplete relevance assessments. Information Retrieval 11(5), 447–470 (2008)

78. Saracevic, T., Kantor, P.: A study of information seeking and retrieving, iii: Searchers, searches, overlap. Journal of the American Society of Information Science 39(3), 197–216 (1988)

79. Savoy, J.: Report on CLEF-2003 Multilingual Tracks. In: Peters, C., Gonzalo, J., Braschler, M., Kluck, M. (eds.) CLEF 2003. LNCS, vol. 3237, pp. 64–73. Springer, Heidelberg (2004)

80. Savoy, J., Calvé, A.L., Vrajitoru, D.: Report on the trec-5 experiment: Data fusion and collection fusion. In: The 5th Text REtrieval Conference (TREC-5), Gaithersburg, USA (November 1996),
http://trec.nist.gov/pubs/trec5/t5_proceedings.html

81. Shokouhi, M.: Central-rank-based collection selection in uncooperative distributed information retrieval. In: Advances in Information Retrieval, Proceedings of the 29th European Conference on IR Research, Rome, Italy, pp. 160–172 (April 2007)

82. Shokouhi, M.: Segmentation of search engine results for effective data-fusion. In: Advances in Information Retrieval, Proceedings of the 29th European Conference on IR Research, Rome, Italy, pp. 185–197 (April 2007)

83. Shokouhi, M., Si, L.: Federated search. Foundations and Trends in Information Retrieval 1(5), 1–102 (2011)

84. Shokouhi, M., Zobel, J.: Robust result merging using sample-based score estimates. ACM Transactions on Information Systems 27(3) (2009)

85. Si, L., Callan, J.: Using sampled data and regression to merge search engine results. In: Proceedings of the 25th Annual International ACM SIGIR Conference, Tampere, Finland, pp. 19–26 (August 2002)

86. Si, L., Callan, J.: Relevant document distribution estimation method for resource selection. In: Proceedings of the 26th Annual International ACM SIGIR Conference, Toronto, Canada, pp. 298–305 (August 2003)

87. Si, L., Callan, J.: A semisupervised learning method to merge search engine results. ACM Transactions on Information Systems 4(21), 457–491 (2003)

88. Soboroff, I., Nicholas, C., Cahan, P.: Ranking retrieval systems without relevance judgments. In: Proceedings of the 24th Annual International ACM SIGIR Conference, New Orleans, Louisiana, USA, pp. 66–73 (September 2001)

89. Spoerri, A.: Examining the authority and ranking effects as the result list depth used in data fusion is varied. Information Processing & Management 4(43), 1044–1058 (2007)

90. Spoerri, A.: Authority and ranking effects in data fusion. Journal of the American Society for Information Science and Technology 3(59), 450–460 (2008)

91. Stock, W.: On relevance distributions. Journal of the American Society for Information Science and Technology 8(57), 1126–1129 (2006)

92. Thewall, M., Hasler, L.: Blog search engines. Online information review 31(4), 467–479 (2007)

93. Thomas, P., Shokouhi, M.: Sushi: Scoring scaled samples for server selection. In: Proceedings of the 32nd Annual International ACM SIGIR Conference, Boston, MA, USA, pp. 419–426 (July 2009)

94. Thompson, P.: Description of the PRC CEO algorithms for TREC. In: The First Text REtrieval Conference (TREC-1), Gaitherburg, MD, USA, pp. 337–342 (March 1993)

95. Tomlinson, S., Oard, D.W., Baron, J.R., Thompson, P.: Overview of the trec 2007 legal track. In: Proceedings of The Sixteenth Text REtrieval Conference, TREC 2007, Gaithersburg, Maryland, USA, November 5-9 (2007), Special Publication 500-274 National Institute of Standards and Technology (NIST)

96. Turtle, H., Croft, W.B.: Evaluation of an inference network-based retrieval model. ACM Transactions on Information Systems 9(3), 187–222 (1991)

97. Vechtomova, O.: Facet-based opinion retrieval from blogs. Information Processing & Management 46(1), 71–88 (2010)

98. Vogt, C.C., Cottrell, G.W.: Predicting the performance of linearly combined IR systems. In: Proceedings of the 21st Annual ACM SIGIR Conference, Melbourne, Australia, pp. 190–196 (August 1998)

99. Vogt, C.C., Cottrell, G.W.: Fusion via a linear combination of scores. Information Retrieval 1(3), 151–173 (1999)

100. Voorhees, E.M., Gupta, N.K., Johnson-Laird, B.: Learning collection fusion strategies. In: Proceedings of the 18th Annual International ACM SIGIR Conference, Seattle, Washington, USA, pp. 172–179 (July 1995)

101. Voorhees, E.M., Harman, D.K. (eds.): Proceedings of the 5th Text Retrieval Conference, Gaithersburg, Maryland, USA. National Institute of Standards and Technology, USA (1996)

102. Wang, Y., Dewitt, D.: Computing pagerank in a distributed internet search engine system. In: Proceedings of the 30th International Conference on Very Large Data Bases, Toronto, Canada, pp. 420–431 (September 2004)

103. John Wilbur, W.: A thematic analysis of the aids literature. In: Proceedings of the 7th Pacific Symposium on Biocomputing, Lihue, Hawaii, USA, pp. 386–397 (January 2002)

104. Wilson, T., Pierce, D.R., Wiebe, J.: Identifying opinionated sentences. In: HLT-NAACL 2003, Human Language Technology Conference of the North American Chapter of the Association for Computational Linguistics, Edmonton, Canada, May 27-June 1 (2003)

105. Wu, S.: A geometric probabilistic framework for data fusion in information retrieval. In: Proceedings of the 10th International Conference on Information Fusion, Quebec, Canada (2007)

106. Wu, S.: Applying statistical principles to data fusion in information retrieval. Expert Systems with Applications 36(2), 2997–3006 (2009)

107. Wu, S.: Applying the data fusion technique to blog opinion retrieval. Expert Systems with Applications 39(1), 1346–1353 (2012)

108. Wu, S.: Linear combination of component results in information retrieval. Data & Knowledge Engineering 71(1), 114–126 (2012)

109. Wu, S., Bi, Y., McClean, S.: Regression relevance models for data fusion. In: Proceedings of the 18th International Workshop on Database and Expert Systems Applications, Regensburg, Germany, pp. 264–268 (September 2007)

110. Wu, S., Bi, Y., Zeng, X.: The Linear Combination Data Fusion Method in Information Retrieval. In: Hameurlain, A., Liddle, S.W., Schewe, K.-D., Zhou, X. (eds.) DEXA 2011, Part II. LNCS, vol. 6861, pp. 219–233. Springer, Heidelberg (2011)

111. Wu, S., Bi, Y., Zeng, X., Han, L.: The Experiments with the Linear Combination Data Fusion Method in Information Retrieval. In: Zhang, Y., Yu, G., Bertino, E., Xu, G. (eds.) APWeb 2008. LNCS, vol. 4976, pp. 432–437. Springer, Heidelberg (2008)

112. Wu, S., Bi, Y., Zeng, X., Han, L.: Assigning appropriate weights for the linear combination data fusion method in information retrieval. Information Processing & Management 45(4), 413–426 (2009)

113. Wu, S., Crestani, F.: Data fusion with estimated weights. In: Proceedings of the 2002 ACM CIKM International Conference on Information and Knowledge Management, McLean, VA, USA, pp. 648–651 (November 2002)

114. Wu, S., Crestani, F.: Distributed information retrieval: A multi-objective resource selection approach. International Journal of Uncertainty, Fuzziness and Knowledge-Based Systems 11(1), 83–100 (2003)

115. Wu, S., Crestani, F.: Methods for ranking information retrieval systems without relevance judgments. In: Proceedings of the 2003 ACM Symposium on Applied Computing (SAC), Melbourne, Florida, USA, pp. 811–816 (March 2003)

116. Wu, S., Crestani, F.: Shadow document methods of results merging. In: Proceedings of the 19th ACM Symposium on Applied Computing, Nicosia, Cyprus, pp. 1067–1072 (March 2004)

117. Wu, S., Crestani, F., Bi, Y.: Evaluating Score Normalization Methods in Data Fusion. In: Ng, H.T., Leong, M.-K., Kan, M.-Y., Ji, D. (eds.) AIRS 2006. LNCS, vol. 4182, pp. 642–648. Springer, Heidelberg (2006)

118. Wu, S., Gibb, F., Crestani, F.: Experiments with document archive size detection. In: Proceedings of the 25th European Conference on Information Retrieval Research, Pisa, Italy, pp. 294–304 (April 2003)

119. Wu, S., McClean, S.: Data fusion with correlation weights. In: Proceedings of the 27th European Conference on Information Retrieval, Santiago de Composite, Spain, pp. 275–286 (March 2005)

120. Wu, S., McClean, S.: Evaluation of System Measures for Incomplete Relevance Judgment in IR. In: Larsen, H.L., Pasi, G., Ortiz-Arroyo, D., Andreasen, T., Christiansen, H. (eds.) FQAS 2006. LNCS (LNAI), vol. 4027, pp. 245–256. Springer, Heidelberg (2006)

121. Wu, S., McClean, S.: Improving high accuracy retrieval by eliminating the uneven correlation effect in data fusion. Journal of American Society for Information Science and Technology 57(14), 1962–1973 (2006)

122. Wu, S., McClean, S.: Information Retrieval Evaluation with Partial Relevance Judgment. In: Bell, D.A., Hong, J. (eds.) BNCOD 2006. LNCS, vol. 4042, pp. 86–93. Springer, Heidelberg (2006)

123. Wu, S., McClean, S.: Performance prediction of data fusion for information retrieval. Information Processing & Management 42(4), 899–915 (2006)

124. Wu, S., McClean, S.: Result merging methods in distributed information retrieval with overlapping databases. Information Retrieval 10(3), 297–319 (2007)

125. Wu, S., Bi, Y., Zeng, X.: Retrieval Result Presentation and Evaluation. In: Bi, Y., Williams, M.-A. (eds.) KSEM 2010. LNCS, vol. 6291, pp. 125–136. Springer, Heidelberg (2010)

126. Wu, S., Crestani, F., Gibb, F.: New Methods of Results Merging for Distributed Information Retrieval. In: Callan, J., Crestani, F., Sanderson, M. (eds.) SIGIR 2003 Ws Distributed IR 2003. LNCS, vol. 2924, pp. 84–100. Springer, Heidelberg (2004)

127. Xu, J., Croft, W.B.: Cluster-based language models for distributed retrieval. In: Proceedings of the 21th Annual International ACM SIGIR Conference, Berkeley, CA, USA, pp. 254–261 (August 1999)

128. Yang, K., Yu, N., Zhang, N.: Widit in trec 2007 blog track: Combining lexicon-based methods to detect opinionated blogs. In: Proceeding of the 16th Text Retrieval Conference, Gaithersburg, MD, USA (2007)

129. Yuwono, B., Lee, D.: Server ranking for distributed text retrieval systems on the internet. In: Proceedings of the Fifth International Conference on Database Systems for Advanced Application, Melbourne, Australia, pp. 41–50 (April 1997)
130. Zhang, W., Yu, C.T., Meng, W.: Opinion retrieval from blogs. In: Proceeding of the 16th ACM Conference on Information and Knowledge Management, Lisbon, Portugal, pp. 831–840 (November 2007)
131. Zipf, G.: Human Behaviour and the principle of least effort: an introduction to human ecology. Addison-Wesley, Reading (1949)
132. Zobel, J.: How reliable are the results of large-scale information retrieval experiments. In: Proceedings of ACM SIGIR Conference, Melbourne, Australia, pp. 307–314 (August 1998)

Appendix A
Systems Used in Section 5.1

Table A.1 TREC 2001 (32 in total, the figures in parentheses are AP values)

apl10wc(0.1567)	apl10wd(0.2035)	flabxt(0.1719)	flabxtd(0.2332)
flabxtdn(0.1843)	flabxtl(0.1705)	fub01be2(0.2225)	fub01idf(0.1900)
fub01ne(0.1790)	fub01ne2(0.1962)	hum01tdlx(0.2201)	iit01tde(0.1791)
jscbtawtl1(0.1890)	jscbtawtl2(0.1954)	jscbtawtl3(0.2003)	jscbtawtl4(0.2060)
kuadhoc2001(0.2088)	Merxtd(0.1729)	msrcn2(0.1863)	msrcn3(0.1779)
msrcn4(0.1878)	ok10wtnd0(0.2512)	ok10wtnd1(0.2831)	pir1Wa(0.1715)
posnir01ptd(0.1877)	ricAP(0.2077)	ricMM(0.2096)	ricMS(0.2068)
ricST(0.1933)	uncfslm(0.0780)	uncvsmm(0.1269)	UniNEn7d(0.2242)

Table A.2 TREC 2003 (62 in total, the figures in parentheses are AP values)

aplrob03a(0.2998)	aplrob03b(0.2522)	aplrob03c(0.2521)
aplrob03d(0.2726)	aplrob03e(0.2535)	fub03IeOLKe3(0.2503)
fub03InB2e3(0.2435)	fub03IneOBu3(0.2329)	fub03IneOLe3(0.2479)
fub03InOLe3(0.2519)	humR03d(0.2367)	humR03de(0.2627)
InexpC2(0.2249)	InexpC2QE(0.2384)	MU03rob01(0.1926)
MU03rob02(0.2187)	MU03rob04(0.2147)	MU03rob05(0.2029)
oce03noXbm(0.2292)	oce03noXbmD(0.1986)	oce03noXpr(0.1853)
oce03Xbm(0.2446)	oce03Xpr(0.1836)	pircRBa1(0.3100)
pircRBa2(0.3111)	pircRBd1(0.2774)	pircRBd2(0.2900)
pircRBd3(0.2816)	rutcor030(0.0482)	rutcor03100(0.0582)
rutcor0325(0.0590)	rutcor0350(0.0683)	rutcor0375(0.0767)
SABIR03BASE(0.2021)	SABIR03BF(0.2263)	SABIR03MERGE(0.2254)
Sel50(0.2190)	Sel50QE(0.2387)	Sel78QE(0.2432)
THUIRr0301(0.2597)	THUIRr0302(0.2666)	THUIRr0303(0.2571)
THUIRr0305(0.2434)	UAmsT03R(0.2324)	UAmsT03RDesc(0.2065)
UAmsT03RFb(0.2452)	UAmsT03RSt(0.2450)	UAmsT03RStFb(0.2373)
UIUC03Rd1(0.2424)	UIUC03Rd2(0.2408)	UIUC03Rd3(0.2502)
UIUC03Rt1(0.2052)	UIUC03Rtd1(0.2660)	uwmtCR0(0.2763)
uwmtCR1(0.2344)	uwmtCR2(0.2692)	uwmtCR4(0.2737)
VTcdhgp1(0.2649)	VTcdhgp3(0.2637)	VTDokrcgp5(0.2563)
VTgpdhgp2(0.2731)	VTgpdhgp4(0.2696)	

Table A.3 TREC 2004 (77 in total, the figures in parentheses are AP values)

apl04rsTDNfw(0.3172)	apl04rsTDNw5(0.2828)	fub04De(0.3062),
fub04Dg(0.3088)	fub04Dge(0.3237)	fub04T2ge(0.2954)
fub04TDNe(0.3391)	fub04TDNg(0.3262)	fub04TDNge(0.3405)
fub04Te(0.2968)	fub04Tg(0.2987)	fub04Tge(0.3089)
humR04d4e5(0.2756)	humR04t5e1(0.2768)	icl04pos2d(0.1746)
icl04pos2f(0.2160)	icl04pos2td(0.1888)	icl04pos48f(0.1825)
icl04pos7f(0.2059)	icl04pos7td(0.1783)	JuruDes(0.2678)
JuruDesAggr(0.2628)	JuruDesLaMd(0.2686)	JuruDesQE(0.2719)
JuruDesSwQE(0.2714)	JuruDesTrSl(0.2658)	JuruTitDes(0.2803)
NLPR04clus10(0.3059)	NLPR04clus9(0.2915)	NLPR04COMB(0.2819)
NLPR04LcA(0.2829)	NLPR04LMts(0.2438)	NLPR04NcA(0.2829)
NLPR04okall(0.2778)	NLPR04OKapi(0.2617)	NLPR04okdiv(0.2729)
NLPR04oktwo(0.2808)	NLPR04SemLM(0.2761)	pircRB04d2(0.3134)
pircRB04d3(0.3319)	pircRB04d4(0.3338)	pircRB04d5(0.3319)
pircRB04t2(0.2984)	pircRB04t3(0.3331)	pircRB04t4(0.3304)
pircRB04td2(0.3586)	pircRB04td3(0.3575)	polyudp2(0.1948)
polyudp4(0.1945)	polyudp5(0.2455)	polyudp6(0.2383)
SABIR04BA(0.2944)	SABIR04BD(0.2627)	SABIR04BT(0.2533)
SABIR04FA(0.2840)	SABIR04FD(0.2609)	SABIR04FT(0.2508)
uogRobDBase(0.2959)	uogRobDWR10(0.3033)	uogRobDWR5(0.3021)
uogRobLBase(0.3056)	uogRobLT(0.3128)	uogRobLWR10(0.3201)
uogRobLWR5(0.3161)	uogRobSBase(0.2955)	uogRobSWR10(0.3011)
uogRobSWR5(0.2982)	vtumdesc(0.2945)	vtumlong252(0.3245)
vtumlong254(0.3252)	vtumlong344(0.3275)	vtumlong348(0.3275)
vtumlong432(0.3275)	vtumlong436(0.3280)	vtumtitle(0.2822)
wdo25qla1(0.2458)	wdoqla1(0.2914)	

Table A.4 TREC 2005 (41 in total, the figures in parentheses are AP values)

DCU05ABM25(0.2887)	DCU05ACOMBO(0.2916)	DCU05ADID(0.2886)
DCU05AWTF(0.3021)	humT05l(0.3154)	humT05x5l(0.3322)
humT05xl(0.3360)	humT05xle(0.3655)	indri05Adm(0.3505)
indri05AdmfL(0.4041)	indri05AdmfS(0.3886)	indri05Aql(0.3252)
JuruDF0(0.2843)	JuruDF1(0.2855)	juruFeRa(0.2752)
juruFeSe(0.2692)	MU05TBa1(0.3199)	MU05TBa2(0.3218)
MU05TBa3(0.3063)	MU05TBa4(0.3092)	NTUAH1(0.3023)
NTUAH2(0.3233)	NTUAH3(0.2425)	NTUAH4(0.2364)
QUT05DBEn(0.1645)	QUT05TBEn(0.1894)	QUT05TBMRel(0.1837)
QUT05TSynEn(0.0881)	sab05tball(0.2087)	sab05tbas(0.2088)
UAmsT05aTeLM(0.1685)	UAmsT05aTeVS(0.1996)	uogTB05LQEV(0.3650)
uogTB05SQE(0.3755)	uogTB05SQEH(0.3548)	uogTB05SQES(0.3687)
uwmtEwtaD00t(0.3173)	uwmtEwtaD02t(0.2173)	uwmtEwtaPt(0.3451)
uwmtEwtaPtdn(0.3480)	york05tAa1(0.1565)	

Appendix B
Systems Used in Section 5.2.4

In every year group, 10 best submission runs are selected. The figures in parentheses are the average performance of that run in RR over 50 queries.

Table B.1 TREC 6

input.uwmt6a0 (0.8770) input.CLAUG (0.8319)
input.CLREL (0.8331)   input.anu6min1 (0.8036)
input.LNmShort (0.8316) input.gerua1 (0.7377)
input.Mercure1 (0.7044) input.anu6alo1 (0.6952)
input.Cor6A3cll (0.6799) input.Brkly22 (0.6776)

Table B.2 TREC 7

input.CLARIT98CLUS (0.8885) input.acsys7mi (0.8324)
input.att98atdc (0.8196)   input.Brkly26 (0.8173)
input.uoftimgr (0.8129)   input.tno7cbm25 (0.8110)
input.tno7exp1 (0.8081)   input.tno7tw4 (0.8058)
input.ok7ax (0.8041)   input.INQ502 (0.7968)

Table B.3 TREC 8

input.manexT3D1N (0.8760) input.CL99SD (0.8588)
input.CL99SDopt1 (0.8853) input.CL99SDopt2 (0.8726)
input.CL99XT (0.8527)   input.CL99XTopt (0.8757)
input.GE8MTD2 (0.8546)   input.Flab8atdn (0.7910)
input.fub99tf (0.7754)   input.apl8p (0.7738)

Table B.4 TREC 8

input.NEnm (0.7133)   input.NEnmLpas (0.7129)
input.iit00td (0.6465)   input.iit00tde (0.6347)
input.Sab9web3 (0.6232) input.tnout9f1 (0.6029)
input.Sab9web2 (0.5862) input.apl9all (0.5814)
input.Sab9web4 (0.5781) input.jscbt9wll2 (0.5681)

Table B.5 TREC 2001

input.ok10wtnd0 (0.7337)   input.ok10wtnd1 (0.7059)
input.kuadhoc2001 (0.6916) input.flabxtd (0.6868)
input.hum01tdlx (0.6736)   input.flabxtdn (0.6627)
input.UniNEn7d (0.6625)   input.ricAP (0.6089)
input.ricMS (0.6064)   input.ricMM (0.6051)

Table B.6 TREC 2002

input.icttd1 (0.4350)   input.Mercah (0.4682)
input.pltr02wt2 (0.4467)   input.thutd1 (0.4630)
input.thutd2 (0.4664)   input.thutd3 (0.4763)
input.thutd4 (0.4613)   input.thutd5 (0.4447)
input.uog03ctadqh (0.4580) input.uog05tad (0.4726)

Appendix C
Systems Used in Sections 7.3-7.5

Table C.1 TREC 8 (the figures in parentheses are values of AP)

input.orcl99man (0.4130)	input.CL99XTopt (0.3766)
input.8manexT3D1N0 (0.3346)	input.pir9Aatd (0.3303)
input.ok8alx (0.3240)	input.Flab8atdn (0.3240)
input.MITSLStdn (0.3227)	input.att99atde (0.3165)
input.apl8p (0.3154)	input.UniNET8Lg (0.3139)

Note: input.READWARE2(0.4632) and input.iit99ma1(0.4104) are not chosen because both of them include far fewer documents (5,785 and 32,061, respectively) than the others (50,000).

Table C.2 TREC 2003 (the figures in parentheses are values of AP)

input.pircRBa2 (0.3111)	input.aplrob03a (0.2998)
input.uwmtCR0 (0.2763)	input.VTgpdhgp2 (0.2731)
input.THUIRr0302 (0.2666)	input.UIUC03Rtd1 (0.2660)
input.humR03de (0.2627)	input.fub03InOLe3 (0.2518)
input.UAmsT03RFb (0.2452)	input.oce03Xbm (0.2446)

Table C.3 TREC 2008 (the figures in parentheses are values of AP)

NOpMMs43 (0.5395)	KGPBASE4 (0.4800)
top3dt1mRd (0.4085)	KLEDocOpinT (0.4062)
B4PsgOpinAZN (0.3924)	DCUCDVPgoo (0.3874)
FIUBL4DFR (0.3760)	b4dt1mRd (0.3742)
DUTIR08Run4 (0.3393)	uams08n1o1sp (0.3351)